Advances in Experimental Medicine and Biology

Volume 1095

More information about this series at http://www.springer.com/series/5584

Heide Schatten

Editor

Cell & Molecular Biology of Prostate Cancer

Updates, Insights and New Frontiers

 Springer

Editor
Heide Schatten
Department of Veterinary Pathobiology
University of Missouri
Columbia, MO, USA

ISSN 0065-2598   ISSN 2214-8019   (electronic)
Advances in Experimental Medicine and Biology
ISBN 978-3-030-07083-0   ISBN 978-3-319-95693-0   (eBook)
https://doi.org/10.1007/978-3-319-95693-0

This Springer imprint is published by the registered company Springer Nature Switzerland AG.
The registered company address is: Gewerbestrasse 11, 6330 Cham, Switzerland

Preface

New advances in cell and molecular biology have led to significant progress in prostate cancer biology, diagnosis, and treatment in which personalized medicine plays an increasingly important role. Basic research, improved imaging modalities, as well as new clinical trials have opened up new avenues to treat this heterogeneous disease with new possibilities of patient-specific approaches. While significant progress has been made in early detection of the disease due to improved diagnostic imaging, treatment of advanced stages of prostate cancer is still in the early stages of research, but progress is being made due to intense efforts to understand cell migration, epithelial-mesenchymal transition points, and metastasis on genetic, cell, and molecular levels that has become possible with newly developed research methods, allowing new insights into the disease.

The advent of molecular technologies has significantly improved our understanding of the biological processes underlying prostate cancer. Targeted therapies are now available to inhibit specific signaling pathways that are aberrant in prostate cancer cell populations, and we are now able to image signaling molecules with specific markers in live cells. Progress has also been made in designing nanoparticles that may be utilized for imaging and targeted prostate cancer treatment. The joint initiatives and efforts of advocate patients, prostate cancer survivors, basic researchers, statisticians, epidemiologists, and clinicians with various and specific expertise have allowed close communication for more specific and targeted treatment. Major forces supporting these efforts are the Department of Defense, the American Cancer Society, and several other foundations that recognized the need for intensified advocacy to find treatments for the disease that represents the most common noncutaneous malignancy for men with new cases resulting in deaths each year.

The present book, *Cell and Molecular Biology of Prostate Cancer: Updates, Insights and New Frontiers*, is one of two companion books; the companion book, *Molecular and Diagnostic Imaging in Prostate Cancer: Clinical Applications and Treatment Strategies*, is focused on clinical aspects. This present book includes classic and modern cell and molecular biology featuring topics to include an overview of prostate cancer statistics, grading, diagnosis, and treatment strategies; novel biomarkers for prostate cancer detection and prognosis; inflammation and prostate cancer; the impact of centrosome pathologies on prostate cancer development and progression; microRNAs as regulators of prostate cancer metastasis; epithelial-mesenchymal transition (EMT) and prostate cancer; the role of multiparametric MRI and fusion biopsy for the diagnosis of prostate cancer; and review of new prostate cancer treatment possibilities.

All articles have been selected as invited chapters written by experts in their specific fields who have made significant contributions to prostate cancer research and present the most recent advances in the field. Cutting-edge new information is balanced with background information that is readily understandable to newcomers and experienced researchers alike. All articles highlight the new aspects of specific research and its impact on designing new strategies or identify new targets for therapeutic intervention. The topics addressed are expected to be of interest to scientists, students, teachers, and to all who are interested in expanding their knowledge related to prostate cancer for diagnostic, therapeutic, or basic research purposes. The books are intended for a large audience as reference books on the subject.

It has been a privilege and great pleasure to edit this volume titled *Cell and Molecular Biology of Prostate Cancer: Updates, Insights and New Frontiers* and the companion book on clinical aspects. I would like to profoundly thank all authors and coauthors for their outstanding contributions and for sharing their unique expertise with the prostate cancer community. I hope the chapters will stimulate further interest in finding new treatment possibilities for this disease to increase the health and survival rates of patients particularly of those suffering from metastatic prostate cancer.

Columbia, MO, USA   Heide Schatten

Contents

Contributors

Divya Bhagirath Department of Urology, Veterans Affairs Medical Center, San Francisco and University of California San Francisco, San Francisco, CA, USA

Rajvir Dahiya Department of Urology, Veterans Affairs Medical Center, San Francisco and University of California San Francisco, San Francisco, CA, USA

Abraham Eisenstark Cancer Research Center, University of Missouri, Columbia, MO, USA

Xavier Filella Department of Biochemistry and Molecular Genetics (CDB), Hospital Clínic, IDIBAPS, Barcelona, Catalonia, Spain

Laura Foj Department of Biochemistry and Molecular Genetics (CDB), Hospital Clínic, IDIBAPS, Barcelona, Catalonia, Spain

Ohuod Hawsawi Department of Biology, Clark Atlanta University, Atlanta, GA, USA

Veronica Henderson Department of Biology, Clark Atlanta University, Atlanta, GA, USA

Irtaza Khan Department of Urology, Icahn School of Medicine at Mount Sinai, New York, NY, USA

Valerie Odero-Marah Department of Biology, Clark Atlanta University, Atlanta, GA, USA

Maureen O. Ripple Dartmouth-Hitchcock/Geisel School of Medicine Office of Development & Geisel School Alumni Relations, Hanover, NH, USA

Sharanjot Saini Department of Urology, Veterans Affairs Medical Center, San Francisco and University of California San Francisco, San Francisco, CA, USA

Debashis Sarkar ST4, Urology Department, St Richard Hospital, Chichester, UK

Heide Schatten Department of Veterinary Pathobiology, University of Missouri, Columbia, MO, USA

Jennifer A. Stockert Department of Urology, Icahn School of Medicine at Mount Sinai, New York, NY, USA

Janae Sweeney Department of Biology, Clark Atlanta University, Atlanta, GA, USA

Ashutosh K. Tewari Department of Urology, Icahn School of Medicine at Mount Sinai, New York, NY, USA

Kamlesh K. Yadav Department of Urology, Icahn School of Medicine at Mount Sinai, New York, NY, USA

Shalini S. Yadav Department of Urology, Icahn School of Medicine at Mount Sinai, New York, NY, USA

Thao Ly Yang Department of Urology, Veterans Affairs Medical Center, San Francisco and University of California San Francisco, San Francisco, CA, USA

Chapter 1
Brief Overview of Prostate Cancer Statistics, Grading, Diagnosis and Treatment Strategies

Heide Schatten

Abstract This chapter provides a brief overview of prostate cancer statistics, grading, diagnosis and treatment strategies that are discussed in more detail in the subsequent chapters of this book and the companion book titled "Clinical Molecular and Diagnostic Imaging of Prostate Cancer and Treatment Strategies". It also points to websites that provide additional useful information for patients affected by prostate cancer and for students and teachers to obtain practical and updated information on research, new diagnostic modalities and new therapies including new updated clinical trials. Three sections are focused on overview of prostate cancer statistics; overview of detection, diagnosis, stages and grading of prostate cancer; and treatment possibilities and options.

Keywords Prostate cancer · Grading · Diagnosis · Treatment · Cytoskeleton · Microtubules · Microfilaments · Metastasis · Chemotherapy

1.1 Introduction

This chapter introduces various aspects of prostate cancer that are discussed in more detail in the subsequent chapters of this book and the companion book titled "Clinical Molecular and Diagnostic Imaging of Prostate Cancer and Treatment Strategies". It also points to websites that provide additional useful information for patients affected by prostate cancer and for students and teachers to obtain practical and updated information on research, new diagnostic modalities and new therapies including new updated clinical trials.

H. Schatten (✉)
Department of Veterinary Pathobiology, University of Missouri, Columbia, MO, USA
e-mail: SchattenH@missouri.edu

© Springer International Publishing AG, part of Springer Nature 2018
H. Schatten (ed.), *Cell & Molecular Biology of Prostate Cancer*,
Advances in Experimental Medicine and Biology 1095,
https://doi.org/10.1007/978-3-319-95693-0_1

1.2   Overview of Prostate Cancer Statistics

Along with skin cancer prostate cancer is the most common cancer in American men. Prostate cancer ranks third in the leading cause of cancer-related deaths in American men behind the first and second leading cancer-causing deaths resulting from lung cancer and colorectal cancer, respectively. It is a disease mostly occurring in older men with 6 cases out of 10 being diagnosed in men aged 65 or older with an average age of about 66 at the time of cancer diagnosis. According to the American Cancer Society (https://www.cancer.org/cancer/prostate-cancer/about/key-statistics.html); (https://cancerstatisticscenter.cancer.org/); (/cancer/prostate-cancer/detection-diagnosis-staging/survival-rates.html) so far for 2017 it is estimated that about 161,360 new cases of prostate cancer will occur with about 26,730 deaths resulting from the disease. Confirmed numbers are available for previous years up to 2014. In 2013, 241,740 new cases of prostate cancer were diagnosed and approximately 28,170 men died from the disease. In 2014, 233,000 new cases with 29,480 deaths resulting from the disease have been reported in the US with similar statistics in other countries.

Based on data obtained for 2010–2014, the number of new cases of prostate cancer was 119.8 per 100,000 men per year and the number of deaths was 20.1 per 100,000 men per year. Based on statistical models for analysis, the rates for new prostate cancer cases have been falling for the past 10 years with death rates falling an average of 3.4% each year which has been attributed to early diagnosis and improved treatment options. Current statistics show that 1 out of 7 men will be diagnosed with prostate cancer during his lifespan and that 1 in 39 men will die of the disease.

Prostate cancer is characterized by abnormally dividing cells in the prostate gland resulting in abnormal prostate gland growth. Most men do not die from prostate cancer but will either be affected by a slow growing tumor or live because of steadily improving and effective treatment. Death from prostate cancer mainly occurs due to metastasis when cancer cells spread to other areas of the body including the pelvic and retroperitoneal lymph nodes, the spinal cord, bladder, rectum, bone and brain.

As our life expectancy has increased significantly over the past few decades it can be expected that the male population with prostate cancer will increase accordingly and it will be important to find new approaches to manage or cure the disease. Currently, there are no sure ways to prevent prostate cancer although some dietary suggestions have been advocated which includes nutrition and lifestyle changes. Most of the research on dietary compounds is still ongoing and clear results and recommendations are not yet available. Some compounds have been proposed to prevent or delay prostate cancer and include extracts from pomegranate, green tea, broccoli, turmeric, crocetin, curcumin, flaxseed and soy among others. Vegan diet (no meat, fish, eggs, or dairy products) and exercise have also been proposed to lower the risks of developing prostate cancer but many of these studies are still not conclusive and the dietary approach and effect may be different for different individuals. Most of these studies including studies of botanical compounds have been tested in animal models to examine the effects on preventing, delaying, or inhibiting prostate cancer but it is not clear how far these results can be extrapolated and applied to human prostate cancer.

The awareness and benefits of early detection and treatment possibilities for prostate cancer has increased success rates to manage or control the disease and new clinical trials are now available to combat the disease in different stages of disease progression. Good progress has been made in developing efficient chemotherapies that will be discussed below in Sect. 1.3.

1.3   Overview of Detection, Diagnosis, Stages and Grading of Prostate Cancer

The various diagnostic methods are only briefly mentioned in this chapter and will be discussed more fully in the individual chapters of this book and the companion book titled "Clinical Molecular and Diagnostic Imaging of Prostate Cancer and Treatment Strategies".

Early Detection Tests Early detection of prostate cancer is important for efficient cures but perfect tests for early detection are not yet reliably available. The prostate-specific antigen (**PSA**) blood test had been used for most previous tests but this test has been critiqued because it may miss some cases of cancer while it may indicate the presence of cancer when prostate cancer could not be found. Aside from the PSA test other tumor markers have been determined (reviewed in detail by [6]) which includes the **phi** test that combines the results of total PSA, free PSA, and proPSA. The **4Kscore** test combines the results of total PSA, free PSA, intact PSA, and human kallikrein 2 (hK2) in addition to other factors that may indicate prostate tumor. The prostate cancer antigen3 (**PCA3**) in the urine is also being assessed after digital rectal exam (DRE), as DRE frees some prostate cells that can be assessed in the urine. More freed prostate cells may indicate a likelihood that prostate cancer is present. Other tests examine abnormal gene changes (**TMPRSS2:ERG**) in cells collected from urine after DRE. This gene is typically not found in men without prostate cancer. **ConfirmMDx** is a test for certain genes in cells obtained from prostate cancer biopsies. These tests may be used to better indicate prostate cancer but more research is needed for more reliable non-invasive tests.

Diagnosis Several methods and technologies are available for diagnosis of prostate cancer. It includes transrectal ultrasound (**TRUS**) to obtain images from areas to be selected for biopsies although TRUS may not reliably detect all areas affected by prostate cancer. Color Doppler ultrasound is an improved technology to detect prostate cancer by measuring blood flow within the gland which may more accurately indicate the areas to be selected for biopsies. This technology has been further improved by employing a contrast agent that can be used to enhance the ultrasound images. Combinations of these technologies can be used for improved detection which includes a combination of MRI with TRUS-guided biopsies.

Stages and Grading of Prostate Cancer Knowing the extent (stage) of prostate cancer is an important factor for determining the treatment options. As new technologies have become available to determine the extent of prostate cancer and new treatment options are also available it has been possible to more accurately diagnose and treat prostate cancer in individuals to determine individualized (personalized) medicine options. Newer methods include multiparametric MRI (reviewed in Sarkar [20]; Schütz et al [31]) to determine the extent and the aggressiveness of prostate cancer and treatment options. This involves a standard MRI and one other type of more detailed MRI such as diffusion weighted imaging (DWI), dynamic contrast enhanced (DCE) MRI or MR spectroscopy.

Another newer method to determine the extent and stage of prostate cancer is enhanced MRI to check lymph nodes for the possibility of containing cancer cells. This method involves a standard MRI followed by an MRI detecting injected magnetic particles.

A newer positron-emission tomography (PET) application involves using radioactive carbon acetate instead of labeled glucose to detect prostate cancer in different parts of the body and to evaluate treatment.

The above mentioned methods are used to determine the stage and possible spreading of cancer. The stages and grades of prostate cancer are described in excellent detail at https://www.cancer.net/cancer-types/prostate-cancer/stages-and-grades which also provides excellent illustrations of the different prostate cancer stages which are briefly described in the following.

Staging and grading of prostate cancer refers to the cancer's growth and spread as well as the particular histology and cellular changes within the tumor. The diagnostic tests described above are used to determine the stages and spread of the cancer. Different grading systems are used for different types of cancer. For prostate cancer, 2 types of staging are used, referred to as the clinical stage and the pathologic stage. The clinical stage is determined by using DRE, biopsy, x-rays, CT and/or MRI scans and bone scans. The pathologic stage refers to information obtained during surgery and test results from the pathology laboratory.

In recent years, the grading system has been redefined based on newly gained results from imaging analysis and newly gained knowledge on prostate cancer. The TNM staging system refers to Tumor (T), Node (N), and Metastasis (M) to address the seizes and location of the tumor (T), spreading of the tumor to lymph nodes (N), and metastasis to other parts of the body (M). The results are then combined to determine the stage of cancer for each individual. Five stages are used to assess the extent of cancer in which 0 refers to no cancer while stages I to IV describe the extent of cancer progression. The details of this grading system are available at the above mentioned website (https://www.cancer.net/cancer-types/prostate-cancer/stages-and-grades) and are not described in this brief overview. Cancer stage grouping is then determined. Stage I describes cancer being confined to the prostate; Stage II describes a tumor in the prostate which is still small and has not spread outside the prostate gland but cells are more abnormal than those found in stage I and cancer has not spread to lymph nodes or distant organs; Stage III refers to the cancer having spread beyond the outer layer of the prostate and can be detected in nearby tissue and also in the seminal vesicles; Stage IV describes a tumor that has spread to other parts of the body, particularly to the bladder, rectum, bone, liver, lungs, or lymph nodes.

Up to recently the Gleason score for grading prostate cancer has mainly been used (reviewed by Giannico and Hameed [9]) and is described as follows. This score is mainly based on morphology/histology/pathology and compares the extent of cancer progression to normal tissue. The Gleason scoring system is the most frequently used grading system and uses a scale of 1 to 5 which determines the pattern of cell growth of the tumor. The specific assessment of cancer cell growth areas are assessed on a scale between 2 and 10 which is then adjusted to the scale of 1–5. Based on the scale between 2 and 10 in recent years physicians are no longer using Gleason scores of 5 or lower for cancers found in biopsies but use 6 as the lowest score to refer to low-grade cancer (reviewed in Giannico and Hameed [9]). A Gleason score of 7 refers to a medium-grade cancer, and a score of 8–10 refers to a high-grade cancer.

On a cellular basis, a Gleason score of x indicates that a Gleason score cannot be determined; a Gleason score of 6 or lower indicates that cells are well differentiated and do not look significantly different than healthy cells; a Gleason score of 7 indicates that cells are moderately differentiated and do not have a pathologic appearance compared to healthy cells; a Gleason score of 8, 9, or 10 indicates that cells are poorly differentiated or undifferentiated and have an abnormal appearance compared to healthy cells. Pathologists have now adopted a Gleason grouping system which simplifies the groups as follows.

Gleason Group I = Former Gleason 6; Gleason Group II = Former Gleason 3 + 4 = 7; Gleason Group III = Former Gleason 4 + 3 = 7; Gleason Group IV = Former Gleason 8; Gleason Group V = Former Gleason 9 or 10.

There are other criteria for staging that have been used by different organizations and are only briefly mentioned here. These relate to risk assessment methods used by the National Comprehensive Cancer Network (NCCN) and the University of San Francisco (UCSF).

The NCCN uses 4 risk-group categories based on PSA level, prostate size, needle biopsy results, and the stage of cancer. The UCSF Cancer of the Prostate Risk Assessment (UCSF-CAPRA) uses a person's age at diagnosis, PSA at diagnosis, Gleason score of the biopsy, T classification from the TNM system, and the percentage of biopsy cores involved with cancer. These criteria are then used in combination to assign a score between 0 and 10 in which a CAPRA score between 0 to 2 indicating low risk, a CAPRA score between 3 to 5 indicating intermediate risk, and a CAPRA score between 6 to 10 indicating high risk.

1.4 Treatment Possibilities and Options

The treatment options are based on the diagnosis that has previously been established by various methods and may include the following. Computerized Tomography (**CT scan**) that uses x-rays to monitor the potential spread of cancer in the body. CT scans will detect cancer that may have spread to lymph nodes or other organs; Magnetic Resonance Imaging (**MRI scan**) is used to image the soft tissue in the body. MRI uses magnets and radio waves instead of x-rays to obtain more detailed images compared to CT scans. It allows imaging of prostate cancer and prostate cancer spread to the prostate-surrounding tissues; Positron Emission Tomography (**PET scan**) uses a tracer liquid to visualize cancer cells. It is often-times employed to find potential cancer remission after treatment; **Lymph node biopsy** is used to determine cancer spread to lymph nodes including lymph nodes in the groin area; **Bone scan** is employed to assess whether or not the cancer has metastasized to bones. For bone scans a low-level radioactive substance is injected to label the bone areas that may be affected by cancer; **Bone biopsy** is performed to assess and confirm results obtained with bone scan. Bone metastasis is among the most frequently observed spreads when prostate cancer metastasizes to different areas in the body. It accounts for about 80 percent of the time when prostate cancer cells metastasize, affecting mostly hip, spine, and pelvis bones. Spreading of pros-tate cancer cells can occur by direct invasion into bones or through the blood or lymphatic system.

Several treatment possibilities are available to control prostate cancer which are discussed in detail in the companion book titled "Clinical Molecular and Diagnostic Imaging of Prostate Cancer and Treatment Strategies". Specific treatments depend on the stages and individual cancer progression. These treatment possibilities are listed in more detail at the American Cancer Society's website (https://www.cancer.org/cancer/prostate-cancer/treating.html) and include the following.

Active Surveillance but no actions are needed (/cancer/prostate-cancer/treating/watchful-waiting.html).

Surgery (/cancer/prostate-cancer/treating/surgery.html). Surgical techniques are constantly improving with the goal to remove all cancer tissue while lowering the risk of complications and side effects resulting from surgery.

Radiation Therapy (/cancer/prostate-cancer/treating/radiation-therapy.html). Radiation therapy has improved significantly in recent years and new technologies are aimed at applying radiation precisely only to the tumor tissue. New technologies include conformal radiation therapy (CRT), intensity modulated radiation therapy (IMRT), and proton beam radiation. As with surgery, the goal is to reduce side effects resulting from radiation therapy. Details for radiation therapy are available in the chapter by Schütz et al. [31].

High-Intensity Focused Ultrasound (HIFU) is a newer treatment procedure used for early stage cancers. It can be used as a first line of therapy or after radiation therapy to treat tissue that has not responded to radiation therapy. HIFU destroys cancer cells by heat using highly focused ultrasonic beams.

Cryotherapy (cryosurgery) (/cancer/prostate-cancer/treating/cryosurgery.html) is also a treatment option.

Hormone Therapy (/cancer/prostate-cancer/treating/hormone-therapy.html). Several new improvements have been made to hormone therapy and include never drugs such as abiraterone and enzalutamine as well as drugs that block the conversion of testosterone to the more active dihydrotestosterone (DHT), 5-alpha reductase inhibitors, such as finasterine and dutasteride. Hormone therapy is further addressed in the text below.

Chemotherapy (/cancer/prostate-cancer/treating/chemotherapy.html) includes taxol in form of paclitaxel or docetaxel and cabazitaxel that are discussed in the text below and target the microtubule system of fast proliferating cancer cells.

Immunotherapy has seen significant progress in recent years and is aimed at boosting the patient's immune system to destroy cancer cells (reviewed in Yadav et al. [32]).

Vaccine Treatment (/cancer/prostate-cancer/treating/vaccine-treatment.html). This prostate cancer treatment (not a prevention treatment) is still limited and currently employs treatment with sipuleucel-T. Several other treatment possibilities in this line of treatments are still in the research phase or in early clinical trials.

Immune Checkpoint Inhibitors are used to prevent cancer cells from disabling the immune system and include drugs such as pembrolizumab and nivolumab that target the immune checkpoint protein PD-1 and lipilimumab that targets the checkpoint protein *CTLA-4* on certain immune cells.

Targeted Therapy Drugs include angiogenesis inhibitors that prevent the growth of new blood vessels to prevent tumor growth. Some angiogenesis inhibitors are currently being tested in clinical trials.

Bone-directed Treatment (/cancer/prostate-cancer/treating/treating-pain.html) uses radiofrequency ablation (RFA) to control metastatic cancer to bones. RFA uses a CT scan or ultrasound to guide a small metal probe into the tumor-affected area, passing a high frequency current through the probe to heat and destroy the tumor.

These treatments can either be applied individually or in combination with other treatments.

Clinical Trials are also available and these are constantly updated as new possibilities become available (/treatment/treatments-and-side-effects/clinical-trials. html).

Basic research has been important for understanding how prostate cancer develops and how abnormalities can be managed. This research has as a goal to understand and increase the treatment options and find new treatment possibilities. Basic research has opened up new directions and new avenues for new treatment possibilities as had been most apparent by the development of taxanes for biomedical research leading to potent treatment of prostate cancer and other cancers. Some of the new basic research has yielded promising translational potential and is advancing into testing in animals and in human clinical trials.

In the early stages of advanced prostate cancer androgen deprivation therapy (ADT) can be successful but most often resistance to androgen deprivation occurs and different treatments are needed which includes taxane-based chemotherapy. We do not yet completely understand the mechanisms leading to resistance to androgen deprivation but it likely involves changes in signal transduction pathways that become aberrant in prostate cancer including the Wnt, PDGF and MAPK pathways. Extensive research on signal transduction pathways has provided some indications for targeted therapies to control prostate cancer cell proliferation (reviewed in Yadav et al. [32]).

The history of prostate cancer treatment has been described by Denmeade and Isaacs [4] and has emphasized the achievements of Huggins and Hodges, who first established the role of male steroid hormones in prostate cancer cell proliferation and the beneficial effects of withdrawal to control prostate tumor growth (reviewed in Martin et al. [16]). Much research has been devoted to the AR which resulted in new cell and molecular data that have led to a better understanding of the effects of androgen withdrawal although we still do not yet clearly understand the pathways involved in AR signaling. It is now known that in the absence of androgen, AR is sequestered in the cytoplasm associated with the Heat Shock Protein 90 (Hsp90) super complex and Filamin A before associating with its ligand, DHT [5, 15]. The structural conformation of the nuclear localization signal (NLS) of the AR does not allow its translocation before binding to DHT [15] but after binding to DHT, AR proteins undergo phosphorylation by Protein Kinase A (PKA) that enables translocation to the nucleus. This process is dependent on the microtubule motor protein dynein that facilitates translocation along microtubules in an ATP- dependent mechanism. This translocation allows binding of the AR to androgen responsive elements (ARE) of the DNA, resulting in proliferation, apoptosis resistance, and epithelial to mesenchymal transition (EMT) (Martin et al. [15]; reviewed in Martin et al. [16]).

The AR is involved in a number of different pathways associated with EMT and metastasis. The pathways involve cadherin switches, Wnt signaling, TGFβ signaling, and Notch signaling (Martin et al. [15]; reviewed in Yadav et al. [32]). During EMT, E-cadherin expression is lost which affects interactions with neighboring cells, as proteins associated with tight junctions are downregulated and cell communication is lost. Loss of cellular communication and subsequent aberrant signal transduction cascades will lead to further metastasis.

When resistance to ADT develops many of the subsequent treatment possibilities involve inhibiting abnormal cytoskeletal dynamics and aberrant cytoskeletal functions to mainly target the microtubule and microfilament activities that are implicated in abnormal cell division and in metastasis to pelvic and retroperitoneal lymph nodes, and to bone [11].

Specific cell and molecular mechanisms that are affected in prostate cancer and specific treatment possibilities are addressed in specific chapters of this book and in the companion book titled "Clinical Molecular and Diagnostic Imaging of Prostate Cancer and Treatment Strategies".

1.4.1 Treatment Possibilities Aimed at Cytoskeletal Abnormalities

The role of the actin and microtubule cytoskeleton in prostate cancer development has been well recognized and drugs targeting their dysfunctions in prostate cancer have been employed successfully with new drugs being developed and tested in laboratory and clinical settings. The cytoskeleton with two of its main components (microtubules and microfilaments) plays a major role in the early stages of prostate cancer initiation leading to abnormal cell proliferation and in cellular mechanisms that allow cancer cells to dissociate from their cellular and tissue organizations to become metastatic. These dissociated cancer cells form seeds to metastasize to different organs, thereby facilitating the epithelial to mesenchymal transition (EMT) (reviewed in detail by [12]). This process includes cells losing their fibroblastic appearance to change their cell shape and become motile. Cell surface changes are significantly associated with changes in the actin cytoskeleton resulting in decreased focal adhesions and downregulation of E-cadherin. Loss of E-cadherin is a critical step in the loss of epidermal adherent junctions that are essential for cells to adhere to each other, allowing cellular communication with neighboring cells and providing cell-cell interactions in normal tissue organizations. Knowing the cell and molecular aspects that are aberrant in cancer development and progression allows the targeted development of therapeutic strategies to eliminate or correct the aberrant processes associated with cancer [23].

As mentioned above, in prostate cancer, the first choice of treatment so far has been the endocrine-targeting approach through androgen deprivation [4] which is highly successful until in many cases tumors become androgen independent and reactivate AR signaling pathways following androgen ablation. The next therapy approach typically employs administration of cytotoxic agents that target the cytoskeleton with the most frequently used microtubule drug taxol [25, 26, 28–30]. While previous studies had used microtubule drugs such as nocodazole or colcemid taxol is unique in that it targets multiple cellular processes to inhibit cell division as well as causing cell destruction. Taxol was isolated from the Pacific Yew tree and modified for the purpose to be used as drug in the 1970's. Taxol binds to microtubules with very high affinity [30], thereby stabilizing microtubules and preventing their dynamic instability that is essential for multiple cellular processes including mitosis and cell division. Paralyzing microtubule functions with taxol results in mitotic block, mitotic cell death and apoptosis [14, 25, 26, 28–30]. Taxol therefore is especially effective in rapidly dividing cancer cells. The specific mechanisms of paclitaxel binding to microtubules have been well studied and discussed above.

Microtubules are highly dynamic cytoskeletal fibers composed of α/β subunit heterodimers that typically are assembled into laterally associated 13 protofilaments to compose one single cylindrical complete microtubule of ca 25 nm diameter. Microtubules display structural polarity characterized by slow growing minus ends and fast growing plus ends. The minus ends can be stabilized by attachment to cellular components such as microtubule organizing centers (MTOCs; centrosomes), the Golgi apparatus, or cell membranes (reviewed in more detail in Schatten and Sun [22, 26]). Individual microtubules undergo phases of growth (polymerization) and shrinkage (depolymerization) in a process termed 'dynamic instability' which allows varied and a great diversity of functions such as forming the mitotic apparatus that separates chromosomes during mitosis and cell division, and a variety of different functions during interphase including maintenance of cell shape, cell motility, cellular transport of membrane vesicles, macromolecules and organelles such as mitochondria.

The role of centrosomes in microtubule organization and functions has been reviewed in several recent papers [21, 22, 24–27] and will be addressed in Chap. 4 of this book.

As mentioned above, several microtubule drugs are known to either inhibit microtubule polymerization (colcemid, colchicine, nocodazole, podophyllotoxin, and griseofulvin) or prevent depolymerization (taxol, paclitaxel). These drugs have different binding properties to microtubules and had been proposed as anticancer drugs to inhibit abnormal cell divisions but taxol has proven the most potent drug that had been identified through basic research [30] and was further developed for clinical applications by investigators at the National Cancer Institute (NCI). Paclitaxel binds to the β subunit of the microtubule (+) end which dimerizes with the α-tubulin subunit [19]. The binding of paclitaxel to the (+) end of microtubules prevents microtubule elongation and prevents microtubule functions. Taxol also blocks cells in the G1 stage of the cell cycle, causing an additional block in interphase added to the block in mitosis.

Microtubule dysfunctions are frequently observed in aging cells and in mitotic cells in which the highly labile microtubules become dysfunctional resulting in spindle abnormalities and aneuploidy. Destabilization of microtubules in aging cells may play a role in the development of age-related cancers and may provide future targets for the prevention of cancer development due to cellular aging.

The actin and microtubule cytoskeletons play a major role in cancer progression with specific roles of the actin cytoskeleton in cellular migration, invasion and metastasis to secondary sites. The actin cytoskeleton consists of its major fiber, the microfilament (F-actin or filamentous actin) composed of its subunits (G-actin or globular actin). Microfilaments consist of a double-helical structure of actin filaments with an intrinsic polarity. One end can rapidly polymerize, termed the plus-end or barbed-end while the other end is the slow growing end called the minus-end or pointed-end.

F- and G-actin interact with a large group of proteins called actin binding proteins (ABPs) [2]. Over 150 ABPs are known to interact with F- and G-actin making up different microfilament organizations to carry out widely different functions. Components of the highly dynamic actin filament system are constantly rearranged and some moving cells form filopodia containing parallel actin filaments that display motility towards an attachment site. Aberrations of regular organizations and functions can lead to various diseases including cancer. In addition, numerous actin-associated proteins are known to play a critical role in regulating actin dynamics and functions. Cellular regulation of the actin cytoskeleton is essential for normal cell function such as cell division, cell locomotion and a great variety of other functions.

Cofilin is an actin binding protein that severs and disassembles actin filaments. Cofilin plays a role in cancer cells and is involved in forming metastatic lesions in patients [33] which is the result of mis-regulation of cofilin. Another prostate cancer promoting factor is TGF-β that functions as tumor growth suppressor in the early stages of cancer, but it becomes activated and enhances cell invasion leading to EMT and metastasis during the late stages of tumor progression (reviewed in Yadav et al. [32]). TGF-β plays a major role in prostate tumor metastasis and invasion; TGF-β and epidermal growth factor (EGF) both stimulate aberrant cofilin expression [34]. In the early stages TGF-β signaling is required to initiate the invasive characteristics towards metastasis [17]. Together with cofilin TGF-β plays a critical role in remodeling of the actin cytoskeleton [2] towards progression to metastasis. The focal adhesion regulator and effector, talin, that acts as an intermediate between integrins and actin is involved in activation of survival pathways towards metastasis.

Although actin and actin-binding proteins can be targeted to arrest cancer cell growth and metastasis based on the significant role in cellular functions and dysfunctions in cancer, so far potent actin-inhibiting chemotherapies aimed at arresting cancer cell proliferation and metastasis have not yet been developed successfully for the clinic to combat prostate cancer although several excellent possibilities have been advanced for translational potential and clinical trials (reviewed in Brayford et al. [1]). So far, the most successful chemotherapeutic drug against cancer cell proliferation and cancer cell destruction remains the microtubule-targeting drug taxol and its new improved derivatives. This is especially true for patients with castration-resistant prostate cancer (CRPC) for which combination therapies are used such as taxanes combined with antiandrogen strategies to increase the survival rate.

Taxane and taxane derivatives inhibit essential microtubule functions and either arrest cells in their cell cycle or cause apoptosis. As mentioned above, microtubules are major components of the mitotic spindle and microtubule dynamics are essential for chromosome separation into the two daughter cells after cell division. Paclitaxel stabilizes microtubules, thereby preventing microtubule dynamics that are essential for spindle functions and chromosome separation, leaving the cell arrested in mitosis and unable to undergo chromosome separation and cell division. Cells in this arrested stage can either undergo mitotic cell death or apoptosis (reviewed in Schatten [21, 25, 26]). Taxol also blocks interphase functions of microtubules that are important for carrying cargo to their functional destinations. This is important for AR translocation, as the N terminal domain of the AR is associated with the α-tubulin subunit of the microtubule. In prostate cancer cells, interphase microtubules play a role in translocation of the androgen receptor (AR) from the cytoplasm to the nucleus. Taxane-induced blockade of microtubule functions can impair androgen receptor activity in prostate cancer by preventing the translocation of the AR into the nucleus, thereby preventing the associated downstream transcriptional activation of AR target genes [3, 35].

Despite the success of employing paclitaxel (taxanes) in prostate cancer chemotherapy cells eventually develop resistance which is mainly due to an affinity of paclitaxel for the overexpressed P-gp (P-glycoprotein) efflux pump that results in loss of effective treatment. For this reason new taxanes have been developed and a second generation taxane, cabazitaxel, is now used, which had been developed based on the rational design of the α-tubulin crystal structure [18]. This new taxane drug carries additional methyl groups that are indirectly attached to ring structures, which allows it to pass the blood brain barrier, thereby affecting tumors metastasizing to the brain which is common for late stage prostate cancer [18].

Other newer microtubule-targeting drugs have been developed which includes the epothilones that also cause microtubule stabilization in similar ways as the taxanes [10]. Unlike taxanes, epothilones and its derivatives similar to cabazitaxel are not affected by the P-glycoprotein efflux pumps, therefore not developing resistance but retaining their cytotoxicity [10, 14] although other types of resistance to epothilones can develop [10].

Other prostate cancer drugs are being developed that target different cellular processes. Novel quinazoline-based compounds, with the lead agent, DZ-50, have been developed as antagonist to the α1-adrenoreceptor [7, 8]. DZ-50 affects metastatic potential in vivo by inhibiting angiogenesis, migration and invasion through targeting focal adhesions [13]. It targets talin and fibronectin in focal adhesion complexes [13]. DZ-50 has now moved into Phase I clinical trials for patients with metastatic CRPC.

1.5 Conclusions and Future Perspective

Prostate cancer is the most common non-cutaneous malignancy for men with new cases resulting in deaths each year. In 2014, 233,000 new cases with 29,480 deaths resulting from the disease have been reported in the US with similar statistics in other countries. Multi-modal approaches are oftentimes required to manage prostate cancer and achieve positive outcomes which requires patient-specific evaluation and analysis for specific management.

New advances in prostate cancer biology have led to significant progress in prostate cancer diagnosis and treatment in which individualized medicine plays an increasingly important role. Basic research, improved imaging modalities as well as new clinical trials has opened up new avenues to treat this heterogeneous disease with new possibilities of patient-specific approaches. While progress has been made in early detection of the disease due to improved diagnostic imaging, treatment of advanced stages of prostate cancer is still in the early stages of research but progress can be foreseen due to intense efforts to understand cell migration, epithelial-mesenchymal transition points, and metastasis on genetic, cell, and molecular levels that has become possible with newly developed research methods.

The advent of molecular technologies has significantly improved our understanding of the biological processes underlying prostate cancer. Targeted therapies are now available to inhibit specific signaling pathways that are aberrant in prostate cancer cell populations and we are now able to image signaling molecules with specific markers in live cells. Progress has also been made in designing nanoparticles that may be utilized for imaging and targeted prostate cancer treatment. The joint initiatives and efforts of advocate patients, prostate cancer survivors, basic researchers, statisticians, epidemiologists, and clinicians with various and specific expertise have allowed close communication for more specific and targeted treatment. Major forces supporting these efforts are the Department of Defense, the American Cancer Society, and several other Foundations that effectively recognized the need for intensified advocacy to find treatments for the disease which has led to falling rates for new prostate cancer death cases to an average of 3.4% for the past 10 years. These efforts are likely to continue due to highly talented and dedicated individuals who are devoted to help combat the disease.

References

1. Brayford S, Schevzov G, Vos J, Gunning P (2015) The role of the actin cytoskeleton in cancer and its potential use as a therapeutic target. In: Schatten H (ed) The cytoskeleton in health and disease. Springer Science+Business Media, New York
2. Collazo J, Zhu B, Larkin S, Martin SK, Pu H, Horbinski C, Koochekpour S, Kyprianou N (2014) Cofilin drives cell-invasive and metastatic responses to TGF-β in prostate cancer. Cancer Res 74(8):2362–2373. https://doi.org/10.1158/0008-5472.CAN-13-3058
3. Darshan MS, Loftus MS, Thadani-Mulero M, Levy BP, Escuin D, Zhou XK, Gjyrezi A, Chanel-Vos C, Shen R, Tagawa ST, Bander NH, Nanus DM, Giannakakou P (2011) Taxane-induced blockade to nuclear accumulation of the androgen receptor predicts clinical responses in metastatic prostate cancer. Cancer Res 71:6019–6029
4. Denmeade SR, Isaacs JT (2002) A history of prostate cancer treatment. Nat Rev Cancer 2:389–396
5. Feldmann BJ, Feldmann D (2001) The development of androgen-independent prostate cancer. Nat Rev Cancer 1:34–45
6. Filella X, Foj L (2018). Novel biomarkers for prostate cancer detection and prognosis. In: Schatten H (ed) Cell and molecular biology of prostate cancer: Updates, insights and new frontiers. Springer Science+Business Media, New York
7. Garrison JB, Kyprianou N (2006) Doxazosin induces apoptosis of benign and malignant prostate cells via death receptor mediated pathway. Cancer Res 66:464–472
8. Garrison JB, Shaw YJ, Chen CS, Kyprianou N (2007) Novel quinazoline-based compounds impair prostate tumorigenesis by targeting tumor vascularity. Cancer Res 67:11344–11352
9. Giannico GA, Hameed O (2018) Evaluation of prostate needle biopsies. In: Schatten H (ed) Clinical molecular and diagnostic imaging of prostate Cancer and treatment strategies. Springer Science Business Media
10. Goodin S, Kane MP, Rubin EH (2004) Epothilones L mechanism of action and biologic activity. J Clin Oncol 22:2015–2025
11. Gravdal K, Halvorsen OJ, Haukaas SA, Akslen LA (2007) A switch from E-cadherin to N-cadherin expression indicates epithelial to mesenchymal transition and is of strong and independent importance for the progress of prostate cancer. Clin Cancer Res 13:7003–7011
12. Hawsawi O, Henderson V, Sweeney J, Odero-Marah V (2018). Epithelial-mesenchymal Transition (EMT) and Prostate Cancer. In: Schatten H (ed) Cell and molecular biology of prostate cancer: Updates, insights and new frontiers. Springer Science+Business Media, New York
13. Hensley PJ, Desinoitis A, Wang C, Stromberg A, Chen CS, Kyprianou N (2014) Novel pharmacalogic targeting of tight junctions and focal adhesions in prostate cancer cells. PLoS One 9:e86238
14. Jordan MA, Wilson L (2004) Microtubules as a target for anticancer drugs. Nat Rev Cancer 4:253–265

15. Martin SK, Fiandalo MV, Kyprianou N (2013) Androgen receptor signaling interactions control epithelial-mesenchymal transition (EMT) in prostate cancer progression. In: editors (ed) Androgen-responsive genes in prostate cancer. Springer, New York, pp 227–255
16. Martin SK, Kamelgarn M, Kyprianou N (2014) Cytoskeleton targeting value in prostate cancer treatment. Am J Clin Exp Urol 2(1):15–26
17. Oft M, Heider KH, Beug H (1998) TGF beta signaling is necessary for carcinoma cell invasiveness and metastasis. Curr Biol 8:1243–1252
18. Paller CJ, Antonarakis ES (2011) Cabazitaxel: a novel second line treatment for metastatic castration-resistant prostate cancer. Drug design. Drug Des Devel Ther 5:117
19. Pellegrini F, Budman DR (2005). Tubulin function, action of antitubulin drugs, and new drug development. Cancer Invest 23(3):264–273.
20. Sarkar D (2018) The role of multi-parametric MRI and fusion biopsy for the diagnosis of prostate cancer. In: Schatten H (ed) Clinical molecular and diagnostic imaging of prostate Cancer and treatment strategies. Springer Science Business Media
21. Schatten H (2008) The mammalian centrosome and its functional significance. Histochem Cell Biol 129:667–686
22. Schatten H (2014) The Role of Centrosomes in Cancer Stem Cell Functions. In: Schatten H (ed). ©Cell and Molecular Biology and Imaging of Stem Cells, First Edition. John Wiley & Sons, Inc, pp 259–279
23. Schatten H (2015) Brief overview of the cytoskeleton. In: Schatten H (ed) The cytoskeleton in health and disease. Springer Science+Business Media, New York
24. Schatten H, Sun Q-Y (2011) The significant role of centrosomes in stem cell division and differentiation. Microsc Microanal 17(4):506–512 Epub 2011 Jul 11
25. Schatten H, Sun QY (2015a) Centrosome and microtubule functions and dysfunctions in meiosis: implications for age-related infertility and developmental disorders. Reprod Fertil Dev 27:934. https://doi.org/10.1071/RD14493. PMID: 25903261
26. Schatten H, Sun Q-Y (2015b) Centrosome-microtubule interactions in health, disease, and disorders. In: Schatten H (ed) The cytoskeleton in health and disease. Springer Science+Business Media, New York
27. Schatten H, Sun Q-Y (2017) Cytoskeletal functions, defects, and dysfunctions affecting human fertilization and embryo development. In: Schatten H (ed) Human reproduction: updates and new horizons. John Wiley & Sons Inc, Hoboken, NJ
28. Schatten G, Schatten H, Bestor T, Balczon R (1982) Taxol inhibits the nuclear movements during fertilization and induces asters in unfertilized sea urchin eggs. J Cell Biol 94:455–465
29. Schatten H, Ripple M, Balczon R, Weindruch R, Taylor M (2000) Androgen and taxol cause cell type specific alterations of centrosome and DNA organization in androgen-responsive LNCaP and androgen-independent prostate cancer cells. Journal of Cellular Biochemistry 76:463–477
30. Schiff PB, Fant J, Horowitz SB (1979) Promotion of microtubule assembly in vitro by taxol. Nature 277:665–667
31. Schütz V, Kesch C, Dieffenbacher S, Bonekamp D, Hadaschik BA, Hohenfellner M, Radtke JP (2018) Multiparametric MRI and MRI/TRUS fusion guided biopsy for the diagnosis of prostate cancer. In: Schatten H (ed) Clinical molecular and diagnostic imaging of prostate Cancer and treatment strategies. Springer Science Business Media
32. Yadav KK, Stockert JA, Yadav SS, Khan I, Tewari AK (2018) Inflammation and prostate cancer. In: Schatten H (ed) Cell and molecular biology of prostate Cancer: updates, insights and new Frontiers. Springer Science Business Media
33. Yoshioka K, Foletta V, Bernard O, Itoh K (2003) A role of LIM kinase in cancer invasion. PNAS 100:7247–7252
34. Zhu B, Fukada K, Zhu H, Kyprianou N (2006) Prohibitin and cofilin are intracellular effectors of transforming growth factor beta signaling in human prostate cancer cells. Cancer Res 66:8640–8647
35. Zhu ML, Horbinski C, Garzotto M, Qian DZ, Beer TM, Kyprianou N (2010) Tubulin targeting chemotherapy impairs androgen receptor activity in prostate cancer. Cancer Res 70:7992–8002

Chapter 2
Novel Biomarkers for Prostate Cancer Detection and Prognosis

Xavier Filella and Laura Foj

Abstract Prostate cancer (PCa) remains as one of the most controversial issues in health care because of the dilemmas related to screening using Prostate Specific Antigen (PSA). A high number of false positive biopsies and an elevated rate of overdiagnosis are the main problems associated with PSA. New PCa biomarkers have been recently proposed to increase the predictive value of PSA. The published results showed that PCA3 score, Prostate Health Index and 4Kscore can reduce the number of unnecessary biopsies, outperforming better than PSA and the percentage of free PSA. Furthermore, 4Kscore provides with high accuracy an individual risk for high-grade PCa. High values of PHI are also associated with tumor aggressiveness. In contrast, the relationship of PCA3 score with aggressiveness remains controversial, with studies showing opposite conclusions. Finally, the development of molecular biology has opened the study of genes, among them TMPRSS2:ERG fusion gene and miRNAs, in PCa detection and prognosis.

Keywords Prostate cancer detection · Biomarker · PSA · Prostate health index 4Kscore · PCA3 score · miRNAs · Exosomal biomarkers

2.1 Introduction

Prostate cancer (PCa) is the second most common cancer in men worldwide, with an estimated 1.1 million new diagnosed cases and 307,000 deaths in 2012 [1]. Furthermore, PCa remains one of the most controversial issues in health care because of the dilemmas related to screening using Prostate Specific Antigen (PSA). PCa detection is difficult due to the limited specificity of PSA, with false positive results in patients with benign prostatic hyperplasia (BPH) as

---

Both authors contributed equally to this manuscript

X. Filella (✉) · L. Foj
Department of Biochemistry and Molecular Genetics (CDB), Hospital Clínic,
IDIBAPS, Barcelona, Catalonia, Spain
e-mail: xfilella@clinic.cat

© Springer International Publishing AG, part of Springer Nature 2018   15
H. Schatten (ed.), *Cell & Molecular Biology of Prostate Cancer*,
Advances in Experimental Medicine and Biology 1095,
https://doi.org/10.1007/978-3-319-95693-0_2

well as in patients with symptomatic and asymptomatic prostatitis. Therefore, biopsy is positive in around 25% of patients with PSA in the range between 2 and 10 μg/L.

PCa is a high prevalent tumor, with an increasing age-related incidence. A systematic review published in 2015 showed that the mean prevalence of incidental PCa in men who died of other causes increased from 5% (95% CI: 3–8%) at age < 30 years to 59% (95% CI:48–71%) by age > 79 years [2]. Therefore, a large proportion of PCa are latent, never progressing into aggressive carcinomas. In this regard, according to the Surveillance, Epidemiology, and End Results, the incidence rate of PCa increased from 94.0 in 1975, in the prePSA era, to 114.14 in 2012, while the death incidence along these years decreased from 30.97 to 19.57 [3]. Actually, PSA screening campaigns cause overdetection of insignificant tumors and thus overtreatment, too. Risks related to overdetection and overtreatment outweigh the potential benefits of screening campaigns.

Currently, PCa guidelines do not recommend the use of PSA as routine test for PCa screening or remark that early PSA testing should be decided considering potential benefits and harms. Debate about the opportunity of screening, nonetheless, goes on. Fleshner et al. [4] recently indicated that the abandonment of PSA screening would prevent all cases of overdiagnosis, but fail to prevent 100% of avoidable deaths, leading to a 13–20% increase in prostate-cancer-related deaths. These data show that harms associated with no screening must also be considered.

PCa is a highly heterogeneous disease in terms of clinical presentation. Different risk classification tools have been developed including biochemical and clinical factors to distinguish patients with PCa according to the prognosis. Epstein criteria [5] have been used to predict insignificant PCa, while the D'Amico classification [6] is used to predict biochemical recurrence after treatment according to biopsy Gleason score, PSA serum levels and the percentage of biopsy material involved with cancer. On the other hand, several authors have put into question if patients with Gleason 6 score must be labeled as cancer, although these tumors have the hallmarks of cancer from a pathologic perspective [7–9]. More recently, new genetic-based evidences confirm the heterogeneity of PCa. The researchers of Cancer Genome Atlas Research Network, analyzing a cohort of 333 tumors, confirmed the molecular heterogeneity of PCa, suggesting a molecular taxonomy in which 74% of PCa tumors are classified in one of seven subtypes defined by specific gene fusions (ERG, ETV1/4, FLI1) or mutations (SPOP, FOXA1, IDH1) [10]. Furthermore, Rubin et al. [11] observed a relationship between genomic amplifications, deletions and point mutations with the prognostic grade groups established in 2016 by the International Society of Urologic Pathology and the World Health Organization to update the Gleason score system.

Active surveillance (AS) has become an alternative to curative therapy for PCa, decreasing the negative effects of overdiagnosis and overtreatment [12]. AS is a way to delay any kind of definitive treatment, applying it only if there is evidence of

progression. The monitoring strategy program includes regular digital rectal examination (DRE), repeated prostate biopsies, and successive measurements of PSA
serum levels to evaluate the PSA doubling time. The selection criteria used to
include patients in an AS program are generally based on D'Amico classification of
low-risk PCa (T1-T2a, PSA < 10 µg/L, Gleason score < 7), although some programs
also include patients with intermediate risk. However, current available criteria to
select patients for AS have a nontrivial risk of misclassification. Therefore, according to a study published by Palisaar et al. [13], current criteria failed around 20% to
identify insignificant PCa from patients who had unfavorable tumor characteristics,
with a high risk of early failure of AS programs and incurable PCa. The availability
of more accurate inclusion criteria would lead to better select patients for AS,
improving the outcome.

New PCa biomarkers have been recently proposed to increase the accuracy
of PSA in the detection and prognosis of early PCa, distinguishing aggressive
and nonaggressive PCa. The search for new subforms of PSA continued in
recent years and new derivatives have been identified. Prostate health index
(PHI) combines [-2]proPSA, free PSA (fPSA) and total PSA, while 4Kscore
-or 4 kallicrein panel- includes total PSA, fPSA, intact PSA (iPSA), and human
kallicrein. Furthermore, the development of molecular biology has opened the
study of genes and the miRNAs associated with PCa. Our aim is to review the
usefulness of these blood and urine new biomarkers in the management of
early PCa.

2.2 PSA–Derived PCa Biomarkers

PSA, also called human kallicrein 3, is a glycoprotein of 30 kDa grouped in the kallicrein family. Because of its enzymatic action, PSA circulates into the blood bound
to several protease inhibitors, such as α-1-antichymotrypsin and α-2-macroglobulin,
whereas only a small fraction, that has been previously inactivated, circulates as free
PSA. The percentage of free PSA to total PSA (%fPSA) is significantly decreased
in patients with PCa, although an overlap of results is observed comparing patients
with and without PCa. Nevertheless, according to a meta-analysis published in
2006, %fPSA only provides additional information in the decision to perform prostate biopsies when levels reach extreme values [14].

The free PSA (fPSA) fraction is also composed of three different subfractions:
benign PSA (BPSA), iPSA, and proPSA. Whereas BPSA is associated with BPH,
proPSA is related to PCa [15]. The native form of proPSA is [-7] proPSA, which
contains a 7-amino acid N-terminal pro-leader peptide. Through the proteolytic
cleavage of this peptide, promoted by the kallikreins hK2 and hK4, the other truncated forms of proPSA, known as [-2] [-4] and [-5] proPSA, are formed.

2.2.1 Prostate Health Index

The truncated forms of proPSA were identified in serum of patients with PCa in 1997, showing that proPSA is a significant fraction of fPSA [16]. Initial published results showed the usefulness of proPSA isoforms in the detection of PCa, reducing the number of negative biopsies in patients with PSA in the grey range. Table 2.1 shows data obtained in initial studies using non-commercial assays for the measurement of one or more isoforms of proPSA [17–21].

Beckman Coulter developed a robust commercial immunoassay for the measurement of [−2]proPSA, or p2PSA. According to results reported by Semjonow et al. [22] p2PSA is stable in serum stored at room temperature or refrigerated at 4 °C for a maximum of 48 h, although blood samples should be centrifuged within 3 h of blood draw. Numerous studies explored the usefulness of p2PSA in the management of early PCa, showing that the percentage of p2PSA in relation to fPSA (%p2PSA) is significantly elevated in patients with PCa. Furthermore, the Prostate Health Index (PHI), a new proPSA derivative indicator, has also yielded promising results in the detection of PCa. This new multiparametric index combines the concentration of p2PSA, fPSA, and total PSA according to the formula (p2PSA/fPSA)* $\sqrt{}$ total PSA.

Both %p2PSA and PHI demonstrated higher accuracy in predicting the presence of PCa at biopsy when compared with total PSA and %fPSA (Table 2.2) [23–27]. Additionally, %p2PSA and PHI showed a good relationship with the aggressiveness of the tumor, with higher levels in patients with Gleason score higher than 6. These

Table 2.1 Summary of studies evaluating proPSA in PCa detection

Authors	Cohort	proPSA isoform	AUCs
Mikolajczyk & Rittenhouse, 2003 [17]	463 patients with PSA 4–10 µg/L	[-2], [-4] & [-5, -7] proPSA	%proPSA/fPSA: 0.689; %fPSA: 0.637; complexed PSA: 0.538
Sokoll et al. 2003 [18]	119 men with PSA 2.5–4 µg/L	[-2], [-4] & [-5, -7] proPSA	%proPSA/fPSA: 0.688; %fPSA: 0.567
Khan et al. 2003 [19]	93 men who underwent a systematic 12-core prostate biopsy (PSA 4–10 µg/L)	[-2], [-4] & [-5, -7] proPSA	%sum proPSA/fPSA: 0.66; total PSA: 0.604; %fPSA: 0.706. Multivariate logistic regression including the sum of proPSA, total PSA and %fPSA: 0.766
Stephan et al. 2006 [20]	1282 patients with PSA 1–10 µg/L	[-5, -7] proPSA	%[-5, -7] proPSA/fPSA: 0.74; %fPSA: 0.73; total PSA: 0.66
Filella et al. 2007 [21]	87 patients with PCa and 138 patients with BPH	[-5, -7] proPSA	%fPSA: 0.705; total PSA: 0.594; multivariate model including [-5, -7] proPSA, %[-5, -7]proPSA/total PSA & %BfPSA/total PSA: 0.753

AUC area under the curve; *PCa* prostate cancer; *BPH* benign prostate hyperplasia; *fPSA* free PSA

Table 2.2 Summary of studies evaluating %p2PSA and PHI in PCa detection

Authors	Cohort	AUCs	Relation of PHI with aggressiveness
Catalona et al. 2011 [23]	892 men with normal DRE, and PSA 2–10 µg/L	PHI: 0.703; %fPSA: 0.648; total PSA: 0.525	Yes, related with Gleason score
Stephan et al. 2013 [24]	1362 patients selected by prostate biopsy with PSA 1.6–8.0 µg/L	PHI:0.74; %p2PSA: 0.72; %fPSA: 0.61; total PSA: 0.56	Yes, related with Gleason score
Lazzeri et al. 2013 [25]	646 patients who underwent prostate biopsy with PSA 2–10 µg/L	PHI: 0.67; %p2PSA: 0.67; %fPSA: 0,64; total PSA: 0.50	Yes, related with Gleason score
Filella et al. 2014 [26]	354 patients with positive or negative prostate biopsy	PHI: 0.732; %p2PSA: 0.723; %fPSA: 0.723; total PSA: 0.553	Yes, related with Gleason score and clinical stage
Loeb et al. 2015 [27]	658 men with PSA 4–10 µg/L and normal DRE who underwent prostate biopsy	PHI: 0.708, %fPSA 0.648, total PSA 0.516	Yes, related with Gleason score and Epstein criteria

AUC area under the curve; *PCa* prostate cancer; *DRE* digital rectal examination; *fPSA* free PSA; *PHI* prostate health index

results are confirmed by three meta-analyses published in 2013 and 2014, which concluded that PHI outperforms the accuracy obtained with PSA and %fPSA, showing an area under the curve (AUC) for PHI from 0.69 to 0.781 [28–30].

Furthermore, the accuracy of PHI in classifying and following patients with PCa on AS has been investigated. Cantiello et al. [31] reported that PHI predicts the pathologic Gleason score, extracapsular extension and seminal vesicles involvement in a series of 156 patients treated with radical prostatectomy. More recently, Heidegger et al. [32] showed that PHI levels were significantly elevated in those patients with an upgrade in final histology (pathologic Gleason score $\geq$ 7) in a cohort of 112 patients with biopsy Gleason score 6 treated with radical prostatectomy. Similarly, De la Calle et al. [33] reported an AUC for PHI of 0.815 to detect high-grade PCa (Gleason score $\geq$ 7). According to these authors, at 95% sensitivity for detecting aggressive PCa the optimal PHI cutoff was 24, which would help to avoid 41% of unnecessary biopsies. On the other hand, baseline and longitudinal %p2PSA and PHI provided improved prediction of biopsy reclassification during follow-up in a series of 167 patients included in a program of active surveillance, according to the results published by Tosoian et al. [34], while total PSA was not significantly associated with biopsy reclassification.

PHI was approved in June 2012 by the US Food and Drug Administration (FDA) for the detection of PCa in men older than 50, PSA between 4 and 10 µg/L, and a non-suspicious DRE. Furthermore, PHI is recommended by the National Comprehensive Cancer Network (NCCN) for patients who have never undergone biopsy or after a negative biopsy. According to NCCN, a PHI higher than 35 is related to a high probability of PCa.

2.2.2 Four–Kallikrein Panel

The four-kallikrein panel includes the measurement of total PSA, fPSA, iPSA and hK2, a protein with high homology to PSA. Several studies performed by the group led by Lilja and Vickers, from Memorial Sloan-Kettering Cancer Center, have evaluated this panel. The AUCs for the 4-kallikrein panel obtained in these studies were higher than those for a PSA based model for the detection of any PCa (AUCs from 0.674 to 0.832) as well as for the detection of high-grade PCa (Gleason score ≥ 7) (AUCs from 0.793 to 0.870) (Table 2.3) [35–39]. Similar results were obtained when DRE was added to those models, showing AUCs from 0.697 to 0.836 in the detection of any PCa and from 0.798 to 0.903 in the detection of high-grade PCa.

The 4Kscore, commercialized by Opko Diagnostics, is an algorithm which combines the four-kallicrein panel with patient age, DRE and history of prior biopsy to predict high-grade PCa. The NCCN Guidelines for PCa recommended the use of 4Kscore for the detection of high-grade tumors. This statistical score improves the specificity for predicting the risk of high-grade PCa, reducing the number of unnecessary biopsies. A prospective study developed in 26 urology centers across the United States, evaluating 1012 men undergoing a prostate biopsy, showed an AUC for 4Kscore of 0.82. The authors reported that 30% of biopsies could be saved using a cut-off value of 6%, delaying diagnosis for 1.3% of high-grade PCa patients [40]. Additionally, Kim et al. [41] recently reported that 4Kscore increased significantly

Table 2.3 Summary of studies evaluating the four-kallikrein panel in the detection of PCa

Authors	Cohort	AUCs Base laboratory model vs. 4 kallikrein panel in detection of PCa	AUCs Base laboratory model vs. 4 kallikrein panel in detection of high-grade PCa[a]
Vickers et al. 2008 [35]	740 unscreened men	0.680 vs. 0.832	0.816 vs. 0.870
Vickers et al. 2010 [36]	2914 unscreened men	0.637 vs. 0.764	0.776 vs. 0.825
Vickers et al. 2010 [37]	1501 previously screened men	0.557 vs. 0.713	0.669 vs. 0.793
Vickers et al. 2010 [38]	1241 men who underwent biopsy for elevated PSA	0.564 vs. 0.674	0.658 vs. 0.819
Vickers et al. 2011 [39]	792 men with PSA ≥ 3 µg/L	0.654 vs. 0.751	0.708 vs. 0.803[b]

AUC area under the curve; *PCa* prostate cancer
Base laboratory model: patient age and total PSA
[a]High-grade cancer was defined as biopsy Gleason score ≥ 7
[b]This study shows AUCs in the detection of palpable PCa, which is defined as clinical stage T2 or higher at diagnosis

the accuracy obtained using the Prostate Cancer Prevention Trial Risk Calculator from 0.73 to 0.79. On the other hand, the test has been shown useful for predicting high-grade PCa in patients with PSA higher than 10 µg/L or with positive DRE, according to a meta-analysis published by Vickers et al. [42]. The addition of the 4Kscore increased the AUC from 0.69 to 0.84 for patients with PSA higher than 10 µg/L and from 0.72 to 0.82 for patients with positive DRE.

Furthermore, Lin et al. [43] showed the ability of 4Kscore to predict high-grade PCa in men included in an AS program. Also, the test has been shown to predict the long term development of distant metastasis. Results published by Stattin et al. [44] showed that the measurement of 4Kscore at 50 and 60 years old allowed the classification of the patients into two groups according to the probability of developing distant metastasis 20 years later. According to this study, patients with 4Kscore higher than 5 at 50 years old and PSA $\geq$ 2 µg/L have a significant increased risk of developing distant metastasis. Also, patients with 4Kscore higher than 7.5 at 60 years old and PSA $\geq$ 3 µg/L have a significant increased risk of developing distant metastasis. The authors concluded that patients with a modest PSA elevation in midlife but a low-risk of high-grade PCa according to 4Kscore could be exempted from biopsy.

2.2.3 PSA Based Nomograms

Several nomograms to predict the likelihood of PCa at biopsy have been developed in last few years with the aim to reduce the number of unnecessary prostate biopsies. These nomograms are graphical representations of a multivariate logistic regression analysis based on specific characteristics of a patient and his disease. Nomograms used to predict PCa combines different demographic, clinical and biochemical variables, including age of the patient, family history of PCa, DRE, prostate volume and PSA serum levels. The Prostate Cancer Risk Calculator of the European Randomized Study of Screening for Prostate Cancer (ERSSPC) (http://www.prostatecancer-riskcalculator.com/) and the Prostate Cancer Prevention Trial (PCPT) based Cancer Risk Calculator (http://myprostatecancerrisk.com/) are among the most used nomograms.

The addition of new biomarkers to these web-based calculators could increase the accuracy for predicting positive prostate biopsies. Lughezzani et al. [45] developed a PHI based nomogram to predict PCa analyzing data from 729 patients who were scheduled for prostate biopsy following suspicious DRE and/or increased PSA. The accuracy increased from 0.73 to 0.80 when PHI was included to a multivariable logistic regression model based on patient age, prostate volume, DRE, and biopsy history. Results were externally validated by a multicenter European study including 833 patients, obtaining an AUC of 0.752 [46]. On the other hand, Filella et al. [47] showed that the accuracy increased from 0.762 to 0.815 when PHI and %p2PSA were added to a multivariable analysis based on patient age, prostate volume, total PSA, and %fPSA. Also, results published by Roobol et al. [48] showed that the addition of PHI to the Prostate Cancer Risk Calculator of the ERSSPC increased the accuracy from 0.65 to 0.72, although it did not increase the accuracy obtained using PHI alone (0.72). Finally, more recently, Loeb et al. [49] reported that adding PHI significantly improved the predictive accuracy of the PCPT and ERSPC risk calculators for aggressive PCa, obtaining an AUC of 0.746.

2.3   mRNA Biomarkers in Urine

Novel mRNA biomarkers have been described in urine, including the mRNAs for PCA3 and TMPRSS2:ERG fusion gene. More recently, positive results have been published for the SelectMDx test, which includes the mRNAs for DLX1, HOXC6 and KLK3 [50, 51].

2.3.1   PCA3

PCA3, previously referred as DD3, is one of the most studied PCa-specific genes, obtaining the FDA's approval in 2012 with the intended use for men older than 50 who have one or more previous negative biopsies. PCA3 is a gene that is overexpressed in PCa tissue [52] and transcribes a long non-coding mRNA involved in PCa cell survival, through modulating the androgen receptor signal [53].

The PCA3 score is calculated as the ratio of PCA3 and PSA mRNAs measured using qRT-PCR in the urine obtained after performing a prostate massage to enrich the prostate cell content. The PSA mRNA is used to normalize the PCA3 mRNA signal and to confirm the specimen validity, controlling the abundance of prostate cells and prostate mRNA. Samples with insufficient PSA mRNA were considered inconclusive. The Progensa PCA3 test, commercialized by Hologic, is a semi-automated assay that includes isolation, amplification, hybridization and quantification of mRNA from PCA3 and PSA using the DTS systems.

A higher PCA3 score is associated with a high prevalence of PCa, improving the results obtained with total PSA [54]. AUCs from 0.63 to 0.87 were documented for PCA3 score in a meta-analysis published in 2010 by Ruiz-Aragón and Márquez-Peláez [55]. Table 2.4 lists similar results reported more recently by other studies [56–60]. The comparative effectiveness review published by Bradley et al. [61] analyzing 34 observational studies showed that PCA3 score is more discriminatory than total PSA, obtaining that at 50% specificity, sensitivities were 77% and 57%, respectively. Nevertheless, differences in accuracy between both tests are lower when the influence of the bias caused by the use of PSA in the selection of patients is minimized. Therefore, Roobol et al. [62] selected patients for biopsy when PSA was 3 μg/L or higher and/or PCA3 score was 10 or higher, showing that PCA3 carries out marginally better than PSA (AUCs of 0.635 and 0.581, respectively; p: 0.143).

Table 2.4 Summary of studies evaluating PCA3 score in the detection of PCa

Authors	Cohort	AUC	Relation of PCA3 with aggressiveness
De la Taille et al. 2011 [56]	515 patients with PSA 2.5–10 µg/L and/ or a suspicious DRE scheduled for initial biopsy	0.761	Yes, with Gleason score
Crawford et al. 2012 [57]	1962 men with PSA > 2.5 µg/L and/or abnormal DRE	0.706	Yes, with Gleason score
Capoluongo et al. 2014 [58]	734 patients who underwent initial prostate biopsy	0.775	No correlation with Gleason score
Chevli et al. 2014 [59]	3073 men who underwent initial prostate biopsy	0.697	No correlation with Gleason score
Foj et al. 2014 [60]	122 patients who underwent prostate biopsy for PSA > 4 µg/L	0.804	No correlation with Gleason score or clinical stage

PCa prostate cancer; AUC area under curve; DRE digital rectal examination

The selection of the most appropriate cut-off for PCA3 remains highly controversial, although 35 is probably the most used cut-off score [55]. The clinical guideline of the NCCN also recommends 35 as the discriminating value to detect PCa, but the FDA suggests that a PCA3 score lower than 25 is associated with a decreased likelihood of a positive biopsy. In this regard, Roobol et al. [62] indicated that 51.7% of biopsies could have been avoided using a cut-off of 35, but the authors underlined that 32% of all PCa and 26.3% of aggressive PCa were missed. Furthermore, according to Bradley et al. [61], the number of missed tumors is reduced significantly from 39% to 6% when the traditional cut-off of 35 is changed for 10, showing that 22% of biopsies were saved using this cut-off.

False positive results are an additional problem using PCA3 score, because a very high PCA3 score does not ensure the existence of PCa. Haese et al. [63] evaluated PCA3 in 463 men with one or two negative biopsies scheduled for repeat biopsy, and found that the probability of a positive repeat prostate biopsy was only of 47% in patients with PCA3 score > 100. Also, Schröder et al. [64] reported a low positive predictive value (38.9%) in a cohort of 56 men with PCA3 score of >100 at previous screens, although significant efforts to detect a PCa were subsequently performed.

Contradictory results have been published by different authors regarding the relationship of PCA3 with the aggressiveness of PCa (Table 2.4). A large study including 3073 men who underwent PCA3 analysis before initial prostate biopsy showed that PCA3 score was significantly associated with biopsy Gleason score [59], although the ROC analysis demonstrated that PCA3 did not significantly outperform PSA in the prediction of high-grade PCa (AUC 0.682 vs. 0.679, respectively, p = 0.702). Furthermore, Auprich et al. [65] showed that PCA3 score failed to add supplementary information to predict aggressive PCa in a series of 305 patients treated with radical prostatectomy, even if the authors obtained a significantly higher median PCA3 in patients with pathological Gleason score 7 or higher. According to these authors, difficulties to pass PCA3 into urine appear in tumors with a high Gleason score because glandular differentiation is lost.

2.3.2 *TMPRSS2:ERG Fusion Gene*

Recurrent chromosomal rearrangements have been observed in several hemato-
logic malignancies and more recently in solid tumors, including PCa.
Approximately 50% of these tumors are associated with fusions involving the
androgen-regulated TMPRSS2 gene with the ETS family transcription factor
family members, particularly ERG and ETV1 [66]. The recent publication of the
Cancer Genome Atlas molecular taxonomy of PCa identifies ETS-rearrangements
as the most common subtype, involving 58% of tumors [10]. These rearrangements result in overexpression of the ETS family transcription factors, which
induces neoplastic phenotype [67]. Furthermore, recent results showed that
TMPRSS2-ERG fusion increases cell migration and promotes cancer metastases in bone [68].

The TMPRSS2:ERG gene rearrangements are detected in urine samples
obtained after a prostate massage using qRT-PCR. Levels of PSA mRNA are
used for control and normalization purposes, and the results are presented as a
TMPRSS2:ERG score. The combination of the TMPRSS2:ERG and PCA3
scores has been proposed as a way to improve the prediction of the presence of
PCa on the biopsy. A recent review underlined that both biomarkers provides
provides 90% specificity and 80% sensitivity in the detection of PCa [69]. A
prospective multicentre evaluation including 443 patients who underwent prostate biopsy underlined that the AUC obtained using the ERSPC risk calculator
increased from 0.799 to 0.842 when PCA3 and TMPRSS2:ERG scores were
added [70]. Additionally, this study reported that TMPRSS2-ERG, but not PCA3,
was associated with the biopsy Gleason score and the tumor clinical stage.
Moreover, the authors found that TMPRSS2-ERG fusion gene was an independent predictor of extracapsular extension of the tumor in a subgroup of 61
patients treated with radical prostatectomy. However, no significant association
was found with pathologic Gleason score or seminal vesicle invasion.

More recently, Tomlins et al. [71] showed the value of PCA3 and
TMPRSS2:ERG scores when they were added to the PCPT risk calculator in a
cohort of 1244 patients. The AUC increased from 0.639 to 0.762 when both tests
were added. Moreover, this study underlined the value of these biomarkers to
predict high-risk PCa, with an AUC of 0.779. Similarly, a recent multicenter
prospective study published by Sanda et al. [72] showed that 42% of unnecessary prostate biopsies would have been avoided combining PCA3 and
TMPRSS2:ERG scores. Furthermore, PCA3 was significantly higher in patients
with Gleason score ≥ 7 versus patients with Gleason score 6. No differences
between both groups of patients were found for TMPRSS2:ERG. These results
were discussed by Stephan et al. [73], who underlined that the combination of
PCA3 and PHI outperformed the accuracy obtained using PCA3 and
TMPRSS2:ERG.

2.4 Exosomal and Non Exosomal miRNAs

2.4.1 MiRNAs Biogenesis, Function and Target Prediction

MicroRNAs (miRNAs) are small, from 18 to 25 nucleotides non-coding RNA molecules that regulate post-transcriptionally gene expression. MiRNAs are derived from so-called pri-miRNA. After being transcribed by RNA polymerase II, pri-miRNA is cleaved by nuclear RNase III Drosha-DGCR8 complex to produce pre-miRNA, which is exported from the nucleus into the cytoplasm by Exportin-5 and Ran-GTP61 and further processed by another endonuclease Dicer to generate mature double-stranded miRNA. Afterwards, the functional strand of the mature miRNA is loaded with Argonaute (AGO) proteins into the RISC (RNA induced silencing complex), where miRNA drives RISC to bind the 3' UTR of a mRNA target, resulting thus in either mRNA cleavage, translational repression or deadenylation. Contrarily, the not functional strand is usually degraded.

Approximately 60% mRNAs can be regulated by miRNAs [74]. Each miRNA can regulate hundreds of genes through base pairing to mRNAs [75]. Moreover, a particular gene can be targeted by multiple miRNAs [76, 77]. Therefore, a miRNA can participate in multiple biological processes by regulating the expression of its target genes [78].

Several tools for target prediction have been developed to understand the molecular mechanisms of miRNA-mediated interactions. Those tools are based on certain assumptions, such as the base complementarity in the 3'UTR, thermodynamic stability, target-site accessibility, and evolutionary conservation of miRNA binding sites. One example of the most used computational prediction is TargetScan [79], which was applied to predict miRNA target sites conserved among orthologous 3' UTRs of vertebrates. However, TargetScan only considers stringent seeds ignoring many potential targets. The intersection of PicTar [80] and TargetScan predictions is recommended in order to achieve both high sensitivity and high specificity.

More recently, Cava et al. [81] described a new software tool, called SpidermiR, which allows to access to both Gene Regulatory Networks and miRNAs in order to obtain miRNA–gene–gene and miRNA–protein–protein interactions. Moreover, SpidermiR integrates this information with differentially expressed genes obtained from The Cancer Genome Atlas through a R/Bioconductor package.

2.4.2 MiRNA in Body Fluids

The last release of miRBase (June, 2014) contains 1881 precursors and 2588 mature human miRNA sequences [82]. The aberrant expression of certain miRNAs has been associated with several cancers including PCa [83]. The dysregulation of miRNAs in cancer could be caused by several genomic anomalies such as chromosomal translocation, epigenetic alterations, as well as miRNA biogenesis machinery dysfunction, which subsequently affects transcription of primary miRNA, its processing to mature miRNAs, and interactions with mRNA targets [84, 85].

Since the initial study of Mitchell et al. [86] in 2008 showing that miRNAs from PCa cells are released into the circulation, different groups have identified several miRNAs signatures with utility in the detection and prognosis of PCa in body fluids (Table 2.5). Specific miRNA signatures in body fluids have been correlated with aggressiveness and response to therapy [86–92]. Nevertheless, there is a lack of concordance across the different studies, probably due to methodological differences that affect several steps of the miRNA analysis from the sample collection to the post-analytical phase. Although the substantial differences among the panels, miR-141, miR-375 and miR-21 are regularly reported in various studies [93].

Mitchell et al. [86] showed that serum levels of miR-141 can distinguish PCa patients from healthy controls, supporting the potential role of this miRNA as a diagnostic marker for PCa. The upregulation of miR-141 in PCa patients was confirmed in later studies [94–97]. At the moment, the widest study about the clinical usefulness of circulating miRNAs has been performed by Mihelich et al. [87], measuring the levels of 21 miRNAs in 50 BPH patients and 100 PCa patients in stages T1–T2, classified according to the Gleason score. High levels of 14 miRNAs were exclusively present in the serum from patients with low-grade PCa or BPH, compared to men with high-grade PCa who had consistently low levels. The expression levels of the 14 miRNAs were combined into a miR Score to predict absence of high-grade PCa among PCa and BPH patients. Furthermore, the authors developed the miR Risk Score based on 7 miRNAs (miR-451, miR-106a, miR-223, miR-107, miR-130b, let-7a and miR-26b) in plasma samples to accurately classify the patients with low-risk of biochemical recurrence. Similarly, Chen et al. [88] found that a panel of five circulating miRNAs (miR-622, miR-1285, let-7e, let-7c, and miR-30c) were significantly different in PCa patients compared to BPH and healthy controls with high accuracy in both identification and validation cohorts. Besides, Cheng et al. [89] identified five serum miRNAs (miR-141, miR-200a, miR-200c, miR-210, and miR-375) associated with metastatic castration resistant PCa. Sharova et al. [90] analysed the levels of circulating miRNAs in patients with elevated PSA who were diagnosed with either localised PCa or BPH upon biopsy and found that miR-106a/miR-130b and miR-106a/miR-223 ratios were significantly different between PCa and BPH groups, concluding that the analysis of the circulating miR-106a/miR-130b (AUC: 0.81) and miR-106a/miR-223 ratios (AUC: 0.77) may reduce the costs and morbidity of unnecessary biopsies. Recently, Al-Qatati et al. [91] using a RT-qPCR based array established a unique expression profile of circulating cell-free miRNAs to differentiate between PCa patients at intermediate versus high-risk for recurrence or death after radical prostatectomy. Particularly, miR-16, miR-148a, and miR-195 were tightly associated with high Gleason score. Those miRNAs are involved in the regulation of the PI3K/Akt signaling pathway and may be promising therapeutic targets for high-risk PCa. Otherwise, Salido-Guadarrama et al. [92] identified a miR-100/200b signature in urine pellet comparing 73 patients with high-risk PCa and 70 patients with BPH. The AUC for this signature (0.738) was higher than the obtained AUCs for total PSA (0.681) and %fPSA (0.710).

Table 2.5 Summary of studies evaluating exosomal and non exosomal miRNAs in body fluids in the detection of PCa

Authors	Body fluid	Patients	Clinical results
Mitchell et al. 2008 [86]	Serum	25 metastatic PCa and 25 matched healthy controls	AUC of 0.907 for miR-141 comparing PCa and healthy controls
Mihelich et al. 2015 [87]	Serum	100 no treated PCa (50 low-grade, 50 high- grade) and 50 BPH	A panel combining let-7a, miR-103, -451, -24, -26b, -30c, -93, -106a, -223, -874, -146a, -125b, -100, -107 and -130b distinguish high-grade PCa from low-grade PCa and BPH
Chen et al. 2012 [88]	Plasma	Screening set: 17 BPH and 25 CaP. Validation set: 44 BPH, 54 healthy controls and 80 CaP	A panel combining miR-622, −1285, −30c, let-7e and let-7c discriminate CaP from BPH (AUC: 0.924) or healthy controls (AUC: 0.860)
Cheng et al. 2013 [89]	Serum	Screening set: 25 mCRPC and 25 age-matched controls. Validation set: 21 mCRPC and 20 age-matched healthy controls	AUCs: miR-141, 0.842; miR-200a, 0.638; miR-200c, 0.645; miR-210, 0.652; miR-375, 0.660 (validation set) miR-210 levels are related to PSA response in mCRPC
Sharova et al. 2016 [90]	Plasma	36 patients with PCa and 31 patients with BPH	miR-106a/miR-130b and miR-106a/miR-223 ratios were significantly different between PCa and BPH groups. AUCs: 0.81 (miR-106a/miR-130b) and 0.77 (miR-106a/miR-223)
Al-Qatati et al. 2017 [91]	Plasma	79 treatment-naïve PCa patients, 1–2 follow-up samples after RP from 51 of the 79 PCa patients, and 33 healthy controls	miR-16, miR-148a and miR-195 significantly correlated with Gleason score. The high miRNA levels before RP remained increased in the postsurgical plasma samples
Salido-Guadarrama et al. 2016 [92]	Urinary pellet	73 patients with HR PCa and 70 patients with BPH	AUC for miR-100/200b signature: 0.738 Adding the miR-100/200b signature to a multivariate model based on age, DRE, total PSA and %fPSA the AUC increased from 0.816 to 0.876
Selth et al. 2012 [94]	Serum	25 mCRPC patients and 25 healthy controls	Levels of miR-141, −298 and − 375 increased in PCa. No correlation with Gleason score, tumor stage, surgical margins, seminal vesicle involvement and extra-capsular extension. Only hsa-miR-298 showed higher expression in tumors with positive surgical margins

(continued)

Table 2.5 (continued)

Authors	Body fluid	Patients	Clinical results
Nguyen et al. 2013 [96]	Serum	28 LR PCa, 30 HR PCa and 26 mCRPC	miR-375, -378*, -141 increase with disease progression
Fredsøe et al. 2017 [98]	Cell-free urine samples	Screening set: 29 BPH patients and 215 patients with clinically localized PCa.	A three-miRNA model (miR-222-3p*miR-24-3p/miR-30c-5p) distinguished BPH and PCa patients with an AUC of 0.95 in screening set, and was successfully validated in validation set (AUC 0.89). Furthermore, a prognostic three-miRNA model (miR-125b-5p*let-7a-5p/miR-151-5p) predicted time to biochemical recurrence after RP in screening set, and was successfully validated
Metcalf et al. 2016 [99]	Serum	16 PValidation set: 29 BPH patients and 220 patients with clinically localized PCa.Ca patients	Elevated levels of miR-141 and miR-375 in patients with active cancers compared to patients in remission, with the highest levels detected in patients with metastatic PCa
Li et al., 2015 [113]	Serum exosomes	Serum vs exosomes cohort: 20 PCa, 20 BPH, 20 healthy controls	Serum exosomal miR-141 was significantly higher in PCa patients compared with BPH patients and healthy controls
Huang et al. 2015 [114]	Plasma exosomes	Screening set: 23 CRPC patients	Plasma exosomal miR-1290 and miR-375 were significantly associated with poor overall survival
		Follow-up set: 100 CRPC patients	
Foj et al. 2017 [115]	Urinary pellets & urinary exosomes	60 PCa patients and 10 healthy controls	A panel combining miR-21 + miR-375 is the best combination to distinguish PCa patients and healthy controls (AUC: 0.872) in urinary pellets.
			MiR-21, -141, -214 were significantly deregulated in intermediate/HR PCa versus LR/healthy subjects in urinary pellets. Significant differences between both groups were found in urinary exosomes for miR-21, -375, and let-7c
Samsonov et al. 2016 [116]	Urinary exosomes	35 PCa patients and 35 healthy controls	miR-21, -141 and - 574 were upregulated in PCa patients compared with healthy controls in urinary exosomes
Alhasan et al. 2016 [118]	Serum	6 VHR PCa patients	AUCs (comparing VHR PCa versus LR PCa and healthy subjects): miR-200c, 1.0; miR-433, 0.98; miR- 135a*, 0.98; miR-605, 0.92; miR-106a: 0.89
		2 HR PCa patients	
		4 LR PCa patients	
		4 healthy controls	

AUC area under the curve; *PCa* prostate cancer; *mCRPC* metastatic castration resistant prostate cancer; *NA* no available; *EV* Extracellular Vesicles; *RP* Radical Prostatectomy; *VHR* very high-risk PCa; *HR* high-risk PCa; *LR* low-risk PCa

Furthermore, when the miR-100/200b signature was included to a multivariate model based on age, DRE, total PSA and %fPSA, the AUC increased from 0.816 to 0.876. Another recent study about miRNAs in urine is the one carried out by Fredsøe et al. [98] who identified several deregulated miRNAs in cell-free urine samples from PCa patients and suggested a novel diagnostic three-miRNA model (miR-222-3p*miR-24-3p/miR-30c-5p) that distinguished BPH and PCa patients with an AUC of 0.95 in a first cohort, and was successfully validated in a second independent cohort (AUC 0.89). Besides, the authors reported a novel prognostic three-miRNA model (miR-125b-5p*let-7a-5p/miR-151-5p) that predicted time to biochemical recurrence after radical prostatectomy independently of routine clinicopathological parameters.

Particularly, Metcalf et al. [99] designed and validated a novel peptide nucleic acids (PNAs) based fluorogenic biosensor for the detection of endogenous concentrations of circulating miRNAs in serum for PCa detection with high affinity and specificity, which does not require any amplification step and involves minimal or no sample processing. The sensing technology is based on oligonucleotide-templated reactions where the only miRNA of interest serves as a matrix to catalyze an otherwise highly unfavorable fluorogenic reaction. The authors used this technology in the serum of 16 PCa patients, finding elevated levels of miR-141 and miR-375 in patients with active cancer compared to patients in remission, with the highest levels detected in metastatic PCa patients. The same RNA samples were analyzed using gold standard RT-qPCR to validate this novel technology and the results were comparable. However, this technology offers the advantages of being low-cost, isothermal and that it is practicable for incorporation into portable devices. Besides, a new feature of this technology is that PNA probes are also capable of detecting miR-precursors, which would indicate that the sum of mature and precursor miRNAs can also be used as a specific biomarker for PCa. Although tests were initially performed on extracted RNA from serum samples, similar results were also obtained when using the probes directly in serum without any amplification and any processing steps.

2.4.3 miRNAs in Exosomes

Exosomes are the smallest (30–150 nm) extracellular vesicles (EV) derived from multivesicular bodies and are involved in intercellular communication, since cells use them to exchange proteins, lipids and nucleic acids [100, 101]. Exosomes are either released from normal or neoplastic cells, being considered to play a fundamental role in many physiological and pathological processes [102]. Exosomes contain mRNA, miRNAs and DNA so the transfer of this kind of information and oncogenic signaling to the tumor microenvironment modulate the tumor progression, the angiogenic proliferation, the formation of the metastasis [103] and even the suppression of immune responses [104].

Several studies have suggested that exosomes obtained from blood and urine are a consistent source of miRNA for disease biomarker detection [105–108], although other researchers highlight that exosomes in standard preparations do not carry biologically significant amount of miRNAs [109]. Actually, RNA sequencing analysis of plasma-derived exosomes revealed that miRNAs are the most abundant exosomal RNA species [100]. The miRNA content of EV reflects the miRNA expression profile of the cells they originated from [110]. Nevertheless, according to Arroyo et al. [111], vesicle associated miRNAs only represents a minority, while around 90% of miRNAs in the circulation is present in a non-membrane-bound form. Contrarily, Gallo et al. [112] showed that the concentration of miRNAs was consistently higher in exosomal fractions as compared to exosome-depleted serum. Cheng et al. [105] performed deep sequencing of miRNAs in exosomal and total cell-free RNA fractions in human plasma and serum and found that exosomes are enriched in miRNAs and provide a consistent source of miRNAs for biomarker discovery. Besides, the same authors found that deep sequencing of exosomal and total cell-free small RNAs in human urine showed a significant enrichment of miRNAs in exosomes.

In fact, few reports have evaluated the exosomal miRNAs utility for PCa detection and prognosis. Li et al. [113] showed that the level of the miR-141 was significantly higher in exosomes compared with whole serum. Besides, according to these authors the level of serum exosomal miR-141 was significantly higher in PCa patients compared with BPH patients and healthy controls, finding the most elevated levels in patients with metastatic PCa. Moreover, Huang et al. [114] found that the levels of plasma exosomal miR-1290 and miR-375 were significantly associated with poor overall survival. The addition of these new biomarkers into a clinical prognostic model improved predictive performance with a time-dependent AUC increase from 0.66 to 0.73. Furthermore, Foj et al. [115] reported that miR-21, miR-375 and let-7c were significantly upregulated in PCa patients versus healthy subjects in urinary exosomes. Additionally, these miRNAs were found significantly deregulated in intermediate/high-risk PCa versus low-risk/healthy subjects in urinary exosomes. Similarly, Samsonov et al. [116] indicated that miR-21, miR-141 and miR-574 were upregulated in PCa patients compared with healthy controls in urinary exosomes isolated by a lectin-based exosomes agglutination method. Nevertheless, only miR-141 was found significantly upregulated when urinary exosomes were isolated by differential centrifugation.

Recently, a high-throughput, spherical nucleic acid-based miRNA expression profiling platform called the Scano-miR bioassay was developed to measure the expression levels of miRNAs with high sensitivity and specificity. The Scano-miR can detect miRNA biomarkers down to 1 femtomolar concentrations and distinguishes perfect miRNA sequences from those with single nucleotide mismatches [117]. Alhasan et al. [118] used the Scano-miR platform to study the exosomal miRNA profiles of serum samples from patients with very high-risk PCa and compared them with the miRNA profiles from healthy individuals and patients with low-risk PCa. The authors identified and validated a unique molecular signature specific for very high-risk PCa. This molecular signature can differentiate patients who may benefit from therapy from those who can be derived to active surveillance. Five miRNA PCa biomarkers (miR-200c, miR-605, miR-135a*, miR-433, and miR-106a) were identified to differentiate low-risk from high and very high-risk PCa.

Due to the discordant results and the lack of overlapping across the different studies, more large-scale studies are needed before clinical application of miRNAs as biomarkers for PCa management. Furthermore, new advances in standardization of all the steps in the process of miRNA analysis are required to improve knowledge on these new biomarkers.

2.5   Conclusion

The published results show that PCA3 score, PHI and 4Kscore can reduce the number of unnecessary biopsies, outperforming better than total PSA and %fPSA. Furthermore, in this review, we have underlined the relationship of these new biomarkers with PCa aggressiveness. The 4Kscore provides with high accuracy an individual risk for high-grade PCa. Also, high values of PHI are associated with tumor aggressiveness. In contrast, the relationship of PCA3 score with aggressiveness remains controversial, with studies showing opposite conclusions. Auprich et al. [119] suggested that the pass of PCA3 into urine is difficult in undifferentiated tumors because of their glandular differentiation lost. In consequence, the PCA3 score measured in urine could be low. On the other hand, more results are necessary to validate the usefulness of TMPRSS2:ERG fusion gene and the exosomal and non-exosomal miRNAs. In addition, new efforts to standardize the methodology used in the measurement of miRNAs are required. Moreover, isolation of exosomes requires easier and more reproducible methods.

Comparison studies among these biomarkers are also necessary to elucidate which of them to select. At the moment, few studies have been reported at this regard, although the performance of PHI and 4Kscore seems similar according to the published results. In this sense, a recent meta-analysis based on twenty-eight studies including 16,762 patients documented comparable AUCs for 4Kscore and PHI in the detection of high-grade PCa (AUCs of 0.81 and 0.82, respectively) [120]. The same conclusion was reported by Nordström et al. [121] evaluating 211 patients undergoing initial or repeat prostate biopsy. The authors showed AUCs in the prediction of high-grade PCa of 0.718 for 4Kscore and 0.711 for PHI.

In contrast, available studies comparing PHI and PCA3 score showed non-conclusive results. Scattoni et al. [122] performed a head-to-head comparison of both biomarkers concluding that PHI was significantly more accurate than the PCA3 score for predicting PCa (AUC 0.70 vs. 0.59). These differences were also observed for predicting PCa in the initial biopsy (AUCs of 0.69 vs 0.57, respectively) and in the repeat biopsy (AUCs of 0.72 vs 0.63, respectively). Conversely, Stephan et al. [123] reported AUCs of 0.74 vs. 0.68 for PCA3 score and PHI, respectively. According to this group, the performance of PCA3 score was slightly better in patients submitted to repeated biopsies than in the first biopsy (AUCs of 0.77 vs 0.70, respectively), while no appreciable differences were reported for PHI between both groups (AUCS of 0.69 vs 0.68, respectively).

To our knowledge, only Vedder et al. [124] performed a study comparing PCA3 score and the 4Kscore. The authors showed that the 4Kscore outperforms the PCA3 score (AUC 0.78 vs. 0.62) in men with elevated PSA, although the accuracy of the PCA3 score was higher in the global population (AUC 0.63 for PCA3 vs. 0.56 for the 4Kscore). Additionally, the authors showed that PCA3 score slightly added value to a multivariate model, increasing the AUC from 0.70 to 0.73. The value of 4Kscore in this regard was minimal, increasing the AUC from 0.70 to 0.71.

Currently, these biomarkers have been recommended by different guidelines [125–127], underlying that they outperform PSA and %fPSA in PCa detection. Furthermore, several studies suggest the value of these biomarkers to increase the predictive accuracy of multivariate models based on classical clinicopathologic variables. Moreover, magnetic resonance imaging (MRI) emerges as a new tool in PCa management, increasing the detection of clinically significant disease. The combined role of these biomarkers together with MRI data should be investigated from an integrated point of view [128]. In summary, available literature shows promising advances in PCa biomarker research. However, large prospective multi-center studies comparing PHI, 4Kscore and PCA3 score are necessary to further elucidate their role in the management of early PCa. Finally, careful validation of emerging biomarkers such as miRNAs and improvement in exosomal isolation are required. The development of these new alternatives could open a new scenario for PCa management in the era of personalized medicine.

References

1. Ferlay J, Steliarova-Foucher E, Lortet-Tieulent J, Rosso S, Coebergh JW, Comber H et al (2013) Cancer incidence and mortality patterns in Europe: estimates for 40 countries in 2012. Eur J Cancer 49:1374–1403
2. Bell KJ, Del Mar C, Wright G, Dickinson J, Glasziou P (2015) Prevalence of incidental prostate cancer: a systematic review of autopsy studies. Int J Cancer 137:1749–1757
3. Howlader N, Noone AM, Krapcho M, et al, eds. (2015) SEER cancer statistics review, 1975–2012, based on November 2014 SEER data submission. National Cancer Institute, Bethesda;. Available at: https://seer.cancer.gov/archive/csr/1975_2012/results_merged/sect_23_prostate.pdf. Accessed 4 July 2017
4. Fleshner K, Carlsson SV, Roobol MJ (2017) The effect of the USPSTF PSA screening recommendation on prostate cancer incidence patterns in the USA. Nat Rev Urol 14:26–37
5. Epstein JI, Walsh PC, Carmichael M, Brendler CB (1994) Pathologic and clinical findings to predict tumor extent of nonpalpable (stage T1c) prostate cancer. JAMA 271:368–374
6. D'Amico AV, Whittington R, Malkowicz SB et al (1998) Biochemical outcome after radical prostatectomy, external beam radiation therapy, or interstitial radiation therapy for clinically localized prostate cancer. JAMA 280:969–974
7. Carter HB, Partin AW, Walsh PC, Trock BJ, Veltri RW, Nelson WG et al (2012) Gleason score 6 adenocarcinoma: should it be labeled as cancer? J Clin Oncol 30:4294–4296

8. Kulac I, Haffner MC, Yegnasubramanian S, Epstein JI, De Marzo AM (2015) Should Gleason 6 be labeled as cancer? Curr Opin Urol 25:238–245
9. Eggener SE, Badani K, Barocas DA, Barrisford GW, Cheng JS, Chin AI et al (2015) Gleason 6 prostate Cancer: translating biology into population health. J Urol 194:626–634
10. Cancer Genome Atlas Research Network (2015) The molecular taxonomy of primary prostate Cancer. Cell 163:1011–1025
11. Rubin MA, Girelli G, Demichelis F (2016) Genomic correlates to the newly proposed grading prognostic groups for prostate Cancer. Eur Urol 69:557–560
12. Tosoian JJ, Carter HB, Lepor A, Loeb S (2016) Active surveillance for prostate cancer: current evidence and contemporary state of practice. Nat Rev Urol 13:205–215
13. Palisaar JR, Noldus J, Löppenberg B, von Bodman C, Sommerer F, Eggert T (2012, Sep) Comprehensive report on prostate cancer misclassification by 16 currently used low-risk and active surveillance criteria. BJU Int 110(6 Pt B):E172–E181
14. Lee R, Localio AR, Armstrong K, Malkowicz SB, Schwartz JS (2006) A meta-analysis of the performance characteristics of the free prostate-specific antigen test. Urology 67:762–768
15. Mikolajczyk SD, Millar LS, Wang TJ et al (2000) A precursor form of prostate-specific antigen is more highly elevated in prostate cancer compared with benign transition zone prostate tissue. Cancer Res 60:756–759
16. Mikolajczyk SD, Grauer LS, Millar LS et al (1997) A precursor form of PSA (pPSA) is a component of the free PSA in prostate cancer serum. Urology 50:710–714
17. Mikolajczyk SD, Rittenhouse HG (2003) Pro PSA: a more cancer specific form of prostate specific antigen for the early detection of prostate cancer. Keio J Med 52:86–91
18. Sokoll LJ, Chan DW, Mikolajczyk SD, Rittenhouse HG, Evans CL, Linton HJ et al (2003) Proenzyme psa for the early detection of prostate cancer in the 2.5-4.0 ng/ml total psa range: preliminary analysis. Urology 61:274–276
19. Khan MA, Partin AW, Rittenhouse HG, Mikolajczyk SD, Sokoll LJ, Chan DW et al (2003) Evaluation of proprostate specific antigen for early detection of prostate cancer in men with a total prostate specific antigen range of 4.0 to 10.0 ng/ml. J Urol 170:723–726
20. Stephan C, Meyer HA, Kwiatkowski M, Recker F, Cammann H, Loening SA et al (2006) A (−5, −7) proPSA based artificial neural network to detect prostate cancer. Eur Urol 50:1014–1020
21. Filella X, Alcover J, Molina R, Luque P, Corral JM, Augé JM et al (2007) Usefulness of proprostate-specific antigen in the diagnosis of prostate cancer. Anticancer Res 27:607–610
22. Semjonow A, Kopke T, Eltze E, Pepping-Schefers B, Burgel H, Darte C (2010) Pre-analytical in-vitro stability of [−2]proPSA in blood and serum. Clin Biochem 43:926–928
23. Catalona WJ, Partin AW, Sanda MG, Wei JT, Klee GG, Bangma CH et al (2011) A multicenter study of [−2]pro-prostate specific antigen combined with prostate specific antigen and free prostate specific antigen for prostate cancer detection in the 2.0 to 10.0 ng/ml prostate specific antigen range. J Urol 185:1650–1655
24. Stephan C, Vincendeau S, Houlgatte A, Cammann H, Jung K, Semjonow A (2013) Multicenter evaluation of [−2]proprostate-specific antigen and the prostate health index for detecting prostate cancer. Clin Chem 59:306–314
25. Lazzeri M, Haese A, de la Taille A, Palou J, McNicholas T, Lughezzani G et al (2013) Serum isoform [−2]proPSA derivatives significantly improve prediction of prostate cancer at initial biopsy in a total PSA range of 2-10 ng/ml: a multicentric European study. Eur Urol 63:986–994
26. Filella X, Foj L, Augé JM, Molina R, Alcover J (2014) Clinical utility of %p2PSA and prostate health index in the detection of prostate cancer. Clin Chem Lab Med 52:1347–1355
27. Loeb S, Sanda MG, Broyles DL, Shin SS, Bangma CH, Wei JT et al (2015) The prostate health index selectively identifies clinically significant prostate cancer. J Urol 193:1163–1169
28. Filella X, Giménez N (2012) Evaluation of [−2]proPSA and prostate health index (phi) for the detection of prostate cancer: a systematic review and meta-analysis. Clin Chem Lab Med 15:1–11

29. Wang W, Wang M, Wang L, Adams TS, Tian Y, Xu J (2014) Diagnostic ability of %p2PSA and prostate health index for aggressive prostate cancer: a meta-analysis. Sci Rep 4:5012
30. Bruzzese D, Mazzarella C, Ferro M, Perdonà S, Chiodini P, Perruolo G et al (2014) Prostate health index vs. percent free prostate-specific antigen for prostate cancer detection in men with "gray" prostate-specific antigen levels at first biopsy: systematic review and meta-analysis. Transl Res 164:444–451
31. Cantiello F, Russo GI, Ferro M, Cicione A, Cimino S, Favilla V et al (2015) Prognostic accuracy of Prostate Health Index and urinary Prostate Cancer Antigen 3 in predicting pathologic features after radical prostatectomy. Urol Oncol 33(163):e15–e23
32. Heidegger I, Klocker H, Pichler R, Pircher A, Prokop W, Steiner E et al (2017 Mar 21) ProPSA and the prostate health index as predictive markers for aggressiveness in low-risk prostate cancer-results from an international multicenter study. Prostate Cancer Prostatic Dis 20:271–275. https://doi.org/10.1038/pcan.2017.3
33. De la Calle C, Patil D, Wei JT, Scherr DS, Sokoll L, Chan DW et al (2015) Multicenter evaluation of the prostate health index to detect aggressive prostate cancer in biopsy naïve men. J Urol 194:65–72
34. Tosoian JJ, Loeb S, Feng Z, Isharwal S, Landis P, Elliot DJ et al (2012) Association of [−2] proPSA with biopsy reclassification during active surveillance for prostate cancer. J Urol 188:1131–1136
35. Vickers AJ, Cronin AM, Aus G, Pihl CG, Becker C, Pettersson K et al (2008) A panel of kallikrein markers can reduce unnecessary biopsy for prostate cancer: data from the European randomized study of prostate Cancer screening in Göteborg. Sweden BMC Med 6:19
36. Vickers A, Cronin A, Roobol M, Savage C, Peltola M, Pettersson K et al (2010) Reducing unnecessary biopsy during prostate cancer screening using a four-kallikrein panel: an independent replication. J Clin Oncol 28:2493–2498
37. Vickers AJ, Cronin AM, Roobol MJ, Savage CJ, Peltola M, Pettersson K et al (2010) A four-kallikrein panel predicts prostate cancer in men with recent screening: data from the European randomized study of screening for prostate Cancer, Rotterdam. Clin Cancer Res 16:3232–3239
38. Vickers AJ, Cronin AM, Aus G, Pihl CG, Becker C, Pettersson K et al (2010) Impact of recent screening on predicting the outcome of prostate cancer biopsy in men with elevated prostate specific antigen: data from the European randomized study of prostate Cancer screening in Gothenburg, Sweden. Cancer 116:2612–2620
39. Vickers AJ, Gupta A, Savage CJ, Pettersson K, Dahlin A, Bjartell A et al (2011) A panel of kallikrein marker predicts prostate cancer in a large, population-based cohort followed for 15 years without screening. Cancer Epidemiol Biomark Prev 20:255–261
40. Parekh DJ, Punnen S, Sjoberg DD, Asroff SW, Bailen JL, Cochran JS et al (2015) A multi-institutional prospective trial in the USA confirms that the 4Kscore accurately identifies men with high-grade prostate cancer. Eur Urol 68:464–470
41. Kim EH, Andriole GL, Crawford ED, Sjoberg DD, Assel M, Vickers AJ et al (2017) Detection of high grade prostate cancer among PLCO participants using a prespecified 4-Kallikrein marker panel. J Urol 197:1041–1047
42. Vickers A, Vertosick EA, Sjoberg DD, Roobol MJ, Hamdy F, Neal D et al (2017) Properties of the 4-Kallikrein panel outside the diagnostic gray zone: meta-analysis of patients with positive digital rectal examination or prostate specific antigen 10 ng/ml and above. J Urol 197(3 Pt 1):607–613
43. Lin DW, Newcomb LF, Brown MD, Sjoberg DD, Dong Y, Brooks JD et al (2016 Nov 23) Evaluating the Four Kallikrein Panel of the 4Kscore for prediction of high-grade prostate cancer in men in the Canary Prostate Active Surveillance Study. Eur Urol pii:S0302-2838(16):30850–30858
44. Stattin P, Vickers AJ, Sjoberg DD, Johansson R, Granfors T, Johansson M et al (2015) Improving the specificity of screening for lethal prostate cancer using prostate-specific antigen and a panel of kallikrein markers: a nested case-control study. Eur Urol 68:207–213

45. Lughezzani G, Lazzeri M, Larcher A, Lista G, Scattoni V, Cestari A et al (2012) Development and internal validation of a prostate health index based nomogram for predicting prostate cancer at extended biopsy. J Urol 188:1144–1150
46. Lughezzani G, Lazzeri M, Haese A, McNicholas T, de la Taille A, Buffi NM et al (2014) Multicenter european external validation of a prostate health index-based nomogram for predicting prostate cancer at extended biopsy. Eur Urol 66:906–912
47. Filella X, Foj L, Alcover J (2014) Aug é JM, Molina R, Jiménez W. The influence of prostate volume in prostate health index performance in patients with total PSA lower than 10 μg/L. Clin Chim Acta 436:303–307
48. Roobol M, Vedder MM, Nieboer D, Houlgatte A, Vincendeau S, Lazzeri M et al (2015) Comparison of two prostate Cancer risk calculators that include the prostate health index. European Urology Focus 1:185–190
49. Loeb S, Shin SS, Broyles DL, Wei JT, Sanda M, Klee G et al (2017) Prostate health index improves multivariable risk prediction of aggressive prostate cancer. BJU Int 120:61–68
50. Leyten GH, Hessels D, Smit FP, Jannink SA, de Jong H, Melchers WJ et al (2015) Identification of a candidate gene panel for the early diagnosis of prostate Cancer. Clin Cancer Res 21:3061–3070
51. Hessels D, de Jong H, Jannink SA, Carter M, Krispin M, Van Criekinge W et al (2017) Analytical validation of an mRNA-based urine test to predict the presence of high-grade prostate cancer. Translational Medicine Communications 2:5. https://doi.org/10.1186/s41231-017-0014-8
52. Bussemakers MJ, van Bokhoven A, Verhaegh GW, Smit FP, Karthaus HF, Schalken JA et al (1999) DD3: a new prostate-specific gene, highly overexpressed in prostate cancer. Cancer Res 59:5975–5979
53. Lemos AE, Ferreira LB, Batoreu NM, de Freitas PP, Bonamino MH, Gimba ER (2016) PCA3 long noncoding RNA modulates the expression of key cancer-related genes in LNCaP prostate cancer cells. Tumour Biol 37:11339–11348
54. Filella X, Foj L, Milà M, Augé JM, Molina R, Jiménez W (2013) PCA3 in the detection and management of early prostate cancer. Tumour Biol 34:1337–1347
55. Ruiz-Aragón J, Márquez-Peláez S (2010) Assessment of the PCA3 test for prostate cancer diagnosis: a systematic review and metaanalysis. Actas Urol Esp 34:346–355
56. De la Taille A, Irani J, Graefen M, Chun F, de Reijke T, Kil P et al (2011) Clinical evaluation of the PCA3 assay in guiding initial biopsy decisions. J Urol 185:2119–2125
57. Crawford ED, Rove KO, Trabulsi EJ, Qian J, Drewnowska KP, Kaminetsky JC et al (2012) Diagnostic performance of PCA3 to detect prostate cancer in men with increased prostate specific antigen: a prospective study of 1962 cases. J Urol 188:1726–1731
58. Capoluongo E, Zambon CF, Basso D, Boccia S, Rocchetti S, Leoncini E et al (2014) PCA3 score of 20 could improve prostate cancer detection: results obtained on 734 Italian individuals. Clin Chim Acta 429:46–50
59. Chevli KK, Duff M, Walter P, Yu C, Capuder B, Elshafei A et al (2014) Urinary PCA3 as a predictor for prostate cancer in a cohort of 3073 men undergoing initial prostate biopsy. J Urol 191:1743–1748
60. Foj L, Milà M, Mengual L, Luque P, Alcaraz A, Jiménez W et al (2014) Real-time PCR PCA3 assay is a useful test measured in urine to improve prostate cancer detection. Clin Chim Acta 435:53–58
61. Bradley LA, Palomaki GE, Gutman S, Samson D, Aronson N (2013) Comparative effectiveness review: prostate cancer antigen 3 testing for the diagnosis and management of prostate cancer. J Urol 190:389–398
62. Roobol MJ, Schröder FH, van Leeuwen P, Wolters T, van den Bergh RC, van Leenders GJ et al (2010) Performance of the prostate cancer antigen 3 (PCA3) gene and prostate-specific antigen in prescreened men: exploring the value of PCA3 for a first-line diagnostic test. Eur Urol 58:475–481

63. Haese A, de la Taille A, van Poppel H, Marberger M, Stenzl A, Mulders PF et al (2008) Clinical utility of the PCA3 urine assay in European men scheduled for repeat biopsy. Eur Urol 54:1081–1088
64. Schröder FH, Venderbos LD, van den Bergh RC, Hessels D, van Leenders GJ, van Leeuwen PJ et al (2014) Prostate cancer antigen 3: diagnostic outcomes in men presenting with urinary prostate cancer antigen 3 scores ≥ 100. Urology 83:613–616
65. Auprich M, Chun FK, Ward JF, Pummer K, Babaian R, Augustin H et al (2011) Critical assessment of preoperative urinary prostate cancer antigen 3 on the accuracy of prostate cancer staging. Eur Urol 59:96–105
66. Tomlins SA, Rhodes DR, Perner S, Dhanasekaran SM, Mehra R, Sun XW et al (2005) Recurrent fusion of TMPRSS2 and ETS transcription factor genes in prostate cancer. Science 310:644–648
67. Deplus R, Delliaux C, Marchand N, Flourens A, Vanpouille N, Leroy X et al (2017) TMPRSS2-ERG fusion promotes prostate cancer metastases in bone. Oncotarget 8:11827–11840
68. Tomlins SA, Laxman B, Dhanasekaran SM, Helgeson BE, Cao X, Morris DS et al (2007) Distinct classes of chromosomal rearrangements create oncogenic ETS gene fusions in prostate cancer. Nature 448:595–599
69. Sanguedolce F, Cormio A, Brunelli M, D'Amuri A, Carrieri G, Bufo P et al (2016) Urine TMPRSS2: ERG fusion transcript as a biomarker for prostate Cancer: literature review. Clin Genitourin Cancer 14:117–121
70. Leyten GH, Hessels D, Jannink SA, Smit FP, de Jong H, Cornel EB et al (2014) Prospective multicentre evaluation of PCA3 and TMPRSS2-ERG gene fusions as diagnostic and prognostic urinary biomarkers for prostate cancer. Eur Urol 65:534–542
71. Tomlins SA, Day JR, Lonigro RJ, Hovelson DH, Siddiqui J, Kunju LP et al (2016) Urine TMPRSS2:ERG plus PCA3 for individualized prostate cancer risk assessment. Eur Urol 70:45–53
72. Sanda MG, Feng Z, Howard DH, Tomlins SA, Sokoll LJ, Chan DW et al (2017 May 18) Association between combined TMPRSS2:ERG and PCA3 RNA urinary testing and detection of aggressive prostate Cancer. JAMA Oncol 3:1085–1093. https://doi.org/10.1001/jamaoncol.2017.0177
73. Stephan C, Cammann H, Jung K (2015) Re: Scott a. Tomlins, John R. Day, Robert J. Lonigro, et al. urine TMPRSS2:ERG plus PCA3 for individualized prostate Cancer risk assessment. Eur Urol 68:e106–e107
74. Friedman RC, Farh KK, Burge CB, Bartel DP (2009) Most mammalian mRNAs are conserved targets of microRNAs. Genome Res 19:92–105
75. Schaefer A, Jung M, Kristiansen G, Lein M, Schrader M, Miller K et al (2010) MicroRNAs and cancer: current state and future perspectives in urologic oncology. Urol Oncol 28:4–13
76. Brennecke J, Stark A, Russell RB, Cohen SM (2005) Principles of microRNA-target recognition. PLoS Biol 3:e85
77. Filipowicz W, Bhattacharyya SN, Sonenberg N (2008) Mechanisms of posttranscriptional regulation by microRNAs: are the answers in sight? Nat Rev Genet 9:102–114
78. Santarpia L, Nicoloso M, Calin GA (2010) MicroRNAs: a complex regulatory network drives the acquisition of malignant cell phenotype. Endocr Relat Cancer 17:51–75
79. Lewis B, Shih I, Jones-Rhoades MW, Bartel DP, Burge CB (2003) Prediction of Mammalian MicroRNA Targets. Cell 115:787–798
80. Krek A, Grün D, Poy MN, Wolf R, Rosenberg L, Epstein EJ et al (2005) Combinatorial microRNA target predictions. Nat Genet 37:495–500
81. Cava C, Colaprico A, Bertoli G, Graudenzi A, Silva TC, Olsen C et al (2017) SpidermiR: an R/Bioconductor package for integrative analysis with miRNA data. Int J Mol Sci 18(2):27. https://doi.org/10.3390/ijms18020274
82. Griffiths-Jones S (2006) miRBase: the microRNA sequence database. Methods Mol Biol Clifton NJ 342:129–138

83. Krol J, Loedige I, Filipowicz W (2010) The widespread regulation of microRNA biogenesis, function and decay. Nat Rev Genet 11:597–610
84. Calin GA, Croce CM (2006) MicroRNA signatures in human cancers. Nat Rev Cancer 6:857–866
85. Decatur WA, Fournier MJ (2002) rRNA modifications and ribosome function. Trends Biochem Sci 27:344–351
86. Mitchell PS, Parkin RK, Kroh EM, Fritz BR, Wyman SK, Pogosova-Agadjanyan EL et al (2008) Circulating microRNAs as stable blood-based markers for cancer detection. Proc Natl Acad Sci U S A 105:10513–10518
87. Mihelich BL, Maranville JC, Nolley R, Peehl DM, Nonn L (2015) Elevated serum microRNA levels associate with absence of high-grade prostate cancer in a retrospective cohort. PLoS One 10:e0124245
88. Chen ZH, Zhang GL, Li HR, Luo JD, Li ZX, Chen GM et al (2012) A panel of five circulating microRNAs as potential biomarkers for prostate cancer. Prostate 72:1443–1452
89. Cheng HH, Mitchell PS, Kroh EM, Dowell AE, Chery L, Siddiqui J et al (2013) Circulating microRNA profiling identifies a subset of metastatic prostate cancer patients with evidence of cancer-associated hypoxia. PLoS One 8:e69239
90. Sharova E, Grassi A, Marcer A, Ruggero K, Pinto F, Bassi P et al (2016) A circulating miRNA assay as a first-line test for prostate cancer screening. Br J Cancer 114:1362–1366
91. Al-Qatati A, Akrong C, Stevic I, Pantel K, Awe J, Saranchuk J et al (2017) Plasma microRNA signature is associated with risk stratification in prostate cancer patients. Int J Cancer 141:1231–1239
92. Salido-Guadarrama AI, Morales-Montor JG, Rangel-Escareño C, Langley E, Peralta-Zaragoza O, Cruz Colin JL et al (2016) Urinary microRNA-based signature improves accuracy of detection of clinically relevant prostate cancer within the prostate-specific antigen grey zone. Mol Med Rep 13:4549–4560
93. Filella X, Foj L (2017) miRNAs as novel biomarkers in the management of prostate cancer. Clin Chem Lab Med 55:715–736
94. Selth LA, Townley S, Gillis JL, Ochnik AM, Murti K, Macfarlane RJ et al (2012) Discovery of circulating microRNAs associated with human prostate cancer using a mouse model of disease. Int J Cancer 131:652–661
95. Brase JC, Johannes M, Schlomm T, Falth M, Haese A, Steuber T et al (2011) Circulating miRNAs are correlated with tumor progression in prostate cancer. Int J Cancer 128:608–616
96. Nguyen HC, Xie W, Yang M, Hsieh CL, Drouin S, Lee GS et al (2013) Expression differences of circulating microRNAs in metastatic castration resistant prostate cancer and low-risk, localized prostate cancer. Prostate 73:346–354
97. Bryant RJ, Pawlowski T, Catto JW, Marsden G, Vessella RL, Rhees B et al (2012) Changes in circulating microRNA levels associated with prostate cancer. Br J Cancer 106:768–774
98. Fredsøe J, Rasmussen AKI, Thomsen AR, Mouritzen P, Høyer S, Borre M et al (2017 Mar 9) Diagnostic and prognostic microRNA biomarkers for prostate cancer in cell-free urine. Eur Urol Focus. https://doi.org/10.1016/j.euf.2017.02.018
99. Metcalf GA, Shibakawa A, Patel H, Sita-Lumsden A, Zivi A, Rama N et al (2016) Amplification-free detection of circulating microRNA biomarkers from body fluids based on fluorogenic oligonucleotide-templated reaction between engineered peptide nucleic acid probes: ppplication to prostate cancer diagnosis. Anal Chem 88:8091–8098
100. Huang X, Yuan T, Tschannen M, Sun Z, Jacob H, Du M et al (2013) Characterization of human plasma-derived exosomal RNAs by deep sequencing. BMC Genomics 14:319
101. Raposo G, Stoorvogel W (2013) Extracellular vesicles: exosomes, microvesicles, and friends. J Cell Biol 200:373–383
102. Théry C, Boussac M, Véron P, Ricciardi-Castagnoli P, Raposo G, Garin J et al (2001) Proteomic analysis of dendritic cell-derived exosomes: a secreted subcellular compartment distinct from apoptotic vesicles. J Immunol 166:7309–7318

103. Simons M, Raposo G (2009) Exosomes–vesicular carriers for intercellular communication. Curr Opin Cell Biol 21:575–581
104. Keller S, Sanderson MP, Stoeck A, Altevogt P (2006) Exosomes: from biogenesis and secretion to biological function. Immunol Lett 107:102–108
105. Cheng L, Sun X, Scicluna BJ, Coleman BM, Hill AF (2014) Characterization and deep sequencing analysis of exosomal and non-exosomal miRNA in human urine. Kidney Int 86:433–444
106. Cheng L, Sharples RA, Scicluna BJ, Hill AF (2014 Mar) Exosomes provide a protective and enriched source of miRNA for biomarker profiling compared to intracellular and cell-free blood. J Extracell Vesicles 26:3. https://doi.org/10.3402/jev.v3.23743
107. Mall C, Rocke DM, Durbin-Johnson B, Weiss RH (2013) Stability of miRNA in human urine supports its biomarker potential. Biomark Med 7:623–631
108. Hessvik NP, Sandvigm K, Llorente A (2013) Exosomal miRNAs as biomarkers for prostate Cancer. Front Genet 4:36
109. Chevillet JR, Kang Q, Ruf IK, Briggs HA, Vojtech LN, Hughes SM et al (2014) Quantitative and stoichiometric analysis of the microRNA content of exosomes. Proc Natl Acad Sci U S A 111:14888–14893
110. Skog J, Würdinger T, van Rijn S, Meijer DH, Gainche L, Sena-Esteves M et al (2008) Glioblastoma microvesicles transport RNA and proteins that promote tumour growth and provide diagnostic biomarkers. Nat Cell Biol 10:1470–1476
111. Arroyo JD, Chevillet JR, Kroh EM, Ruf IK, Pritchard CC, Gibson DF et al (2011) Argonaute2 complexes carry a population of circulating microRNAs independent of vesicles in human plasma. Proc Natl Acad Sci U S A 108:5003–5008
112. Gallo A, Tandon M, Alevizos I, Illei GG (2012) The majority of microRNAs detectable in serum and saliva is concentrated in exosomes. PLoS One 7:e30679
113. Li Z, Ma YY, Wang J, Zeng XF, Li R, Kang W et al (2015) Exosomal microRNA-141 is upregulated in the serum of prostate cancer patients. Onco Targets Ther 9:139–148
114. Huang X, Yuan T, Liang M, Du M, Xia S, Dittmar R et al (2015) Exosomal miR-1290 and miR-375 as prognostic markers in castration-resistant prostate cancer. Eur Urol 67:33–41
115. Foj L, Ferrer F, Serra M, Arévalo A, Gavagnach M, Giménez N et al (2017) Exosomal and non-Exosomal urinary miRNAs in prostate Cancer detection and prognosis. Prostate 77:573–583
116. Samsonov R, Shtam T, Burdakov V, Glotov A, Tsyrlina E, Berstein L et al (2016) Lectin-induced agglutination method of urinary exosomes isolation followed by mi-RNA analysis: application for prostate cancer diagnostic. Prostate 76:68–79
117. Alhasan AH, Kim DY, Daniel WL, Watson E, Meeks JJ, Thaxton CS et al (2012) Scanometric microRNA array profiling of prostate cancer markers using spherical nucleic acid-gold nanoparticle conjugates. Anal Chem 84:4153–4160
118. Alhasan AH, Scott AW, Wu JJ, Feng G, Meeks JJ, Thaxton CS et al (2016) Circulating microRNA signature for the diagnosis of very high-risk prostate cancer. Proc Natl Acad Sci U S A 113:10655–10660
119. Auprich M, Bjartell A, Chun FK, de la Taille A, Freedland SJ, Haese A et al (2011) Contemporary role of prostate cancer antigen 3 in the management of prostate cancer. Eur Urol 60:1045–1054
120. Russo GI, Regis F, Castelli T, Favilla V, Privitera S, Giardina R et al (2017) A systematic review and meta-analysis of the diagnostic accuracy of prostate health index and 4-kallikrein panel score in predicting overall and high-grade prostate cancer. Clin Genitourin Cancer 15:429–439
121. Nordström T, Vickers A, Assel M, Lilja H, Grönberg H, Eklund M (2015) Comparison between the four-kallikrein panel and prostate health index for predicting prostate Cancer. Eur Urol 68:139–146

122. Scattoni V, Lazzeri M, Lughezzani G, de Luca S, Passera R, Bollito E et al (2013) Head-to-head comparison of prostate health index and urinary PCA3 for predicting cancer at initial or repeat biopsy. J Urol 190:496–501
123. Stephan C, Jung K, Semjonow A, Schulze-Forster K, Cammann H, Hu X et al (2013) Comparative assessment of urinary prostate cancer antigen 3 and TMPRSS2:ERG gene fusion with the serum [−2]proprostate-specific antigen-based prostate health index for detection of prostate cancer. Clin Chem 59:280–288
124. Vedder MM, de Bekker-Grob EW, Lilja HG, Vickers AJ, van Leenders GJ, Steyerberg EW et al (2014) The added value of percentage of free to total prostate-specific antigen, PCA3, and a kallicrein panel to the ERSPC risk calculator for prostate Cancer in prescreened men. Eur Urol 66:1109–1115
125. Prostate Cancer Early Detection (2017) National Cancer Comprehensive Network Clinical Practice Guidelines in Oncology. Version I. 2017. Available online: https://www.nccn.org/professionals/physician_gls/pdf/prostate_detection.pdf (Accessed on 18 August 2017)
126. Mottet N, Bellmunt J, Briers E, Bolla M, Cornford P, de Santis M, et al. (n.d.) EAU Guideliness prostate cancer. Available online: http://uroweb.org/guideline/prostate-cancer/ (Accessed on 18 August 2017)
127. Vickers AJ, Eastham JA, Scardino PT, Lilja H (2016) The memorial Sloan Kettering Cancer center recommendations for prostate Cancer screening. Urology 91:12–18
128. Watson MJ, George AK, Maruf M, Frye TP, Muthigi A, Kongnyuy M et al (2016) Risk stratification of prostate cancer: integrating multiparametric MRI, nomograms and biomarkers. Future Oncol 12:2417–2430

Chapter 3
Inflammation and Prostate Cancer

Ashutosh K. Tewari, Jennifer A. Stockert, Shalini S. Yadav, Kamlesh K. Yadav, and Irtaza Khan

Abstract Chronic inflammation resulting from infections, altered metabolism, inflammatory diseases or other environmental factors can be a major contributor to the development of several types of cancer. In fact around 20% of all cancers are linked to some form of inflammation. Evidence gathered from genetic, epidemiological and molecular pathological studies suggest that inflammation plays a crucial role at various stages of prostatic carcinogenesis and tumor progression. These include initiation, promotion, malignant conversion, invasion, and metastasis. Detailed basic and clinical research in these areas, focused towards understanding the etiology of prostatic inflammation, as well as the exact roles that various signaling pathways play in promoting tumor growth, is critical for understanding this complex process. The information gained would be useful in developing novel therapeutic strategies such as molecular targeting of inflammatory mediators and immunotherapy-based approaches.

Keywords Prostate cancer · Inflammation · Oxidative stress · Innate immune system · Adaptive immune system · Macrophages · Lymphocytes · Cytokines Chemokines

Abbreviations

AML	Acute myeloid leukemia
ARE	Antioxidant response element
Bcl	B-cell lymphoma
BMI1	B lymphoma Mo-MLV insertion region 1 homolog
BMP7	Bone morphogenetic protein 7
BPH	Benign prostatic hyperplasia

A. K. Tewari (✉) · J. A. Stockert · S. S. Yadav · K. K. Yadav · I. Khan
Department of Urology, Icahn School of Medicine at Mount Sinai, New York, USA
e-mail: ash.tewari@mountsinai.org

© Springer International Publishing AG, part of Springer Nature 2018   41
H. Schatten (ed.), *Cell & Molecular Biology of Prostate Cancer*,
Advances in Experimental Medicine and Biology 1095,
https://doi.org/10.1007/978-3-319-95693-0_3

BRCA1	Breast cancer 1
CCL2	C-C motif chemokine ligand 2
CD	Cluster of differentiation
CDKN1B	Cyclin dependent kinase inhibitor 1B
CMV	Cytomegalovirus
COX2	Cyclooxygenase-2
CSC	Cancer stem cells
CSF-1R	Colony stimulating factor 1 receptor
CXCL12	C-X-C motif chemokine ligand 12
DAMP	Danger-associated molecular patterns
DHT	Dihydrotestosterone
DNA	Deoxyribonucleic acid
DNMT1	DNA methyltransferase 1
DPI	Diphenyleneiodonium
EBV	Epstein-Barr virus
EMT	Epithelial to mesenchymal transformation
EZH2	Enhancer of zeste homology 2
GATA1	GATA sequence binding factor 1
GST	Glutathione-S-transferase
GSTP1	Glutathione-S-transferase Pi 1
HCA	Heterocyclic amines
HCC	Hepatocellular carcinoma
Hh	Hedgehog
HMG-CoA	3-hydroxy-3-methylglutaryl-coenzyme A
HNE	4-hydroxy-2-nonenal
HPV	Human papillomavirus
HSV2	Herpes simplex virus 2
IFN	Interferon
IL	Interleukin
JAK	Janus kinase
LNCaP	Prostate adenocarcinoma cell line
MAPK	Mitogen activated protein kinase
mCRPC	Metastatic castration-resistant prostate cancer
M-CSF	Macrophage colony stimulating factor
MDSC	Myeloid-derived suppressor cells
MIC1	Macrophage inhibitory cytokine 1
miR	MicroRNA
MLH1	MutL homolog 1
MMPs	Matrix metalloproteinases
MSR1	Macrophage scavenger receptor 1
NADPH	Nicotinamide adenine dinucleotide phosphate
NF-κB	Nuclear factor kappa B subunit
NKX3.1	NK3 transcription factor related, locus 1
NO	Nitric oxide
Nrf2	Nuclear factor erythroid 2-related factor 2

NSAID	Nonsteroidal anti-inflammatory drug
PAMP	Pathogen-associated molecular patterns
PC3	Prostate CRPC cell line
PCa	Prostate cancer
PI3K	Phosphoinositide 3-kinase
PIA	Proliferative inflammatory atrophy
PIN	Prostatic intraepithelial neoplasia
PRR	Pattern recognition receptors
PSA	Prostate specific antigen
RNS	Reactive nitrogen species
ROS	Reactive oxen species
RUNX3	Runt related transcription factor 3
SMAD	Small body size mothers against decapentaplegic
STAT3	Signal transducer and activator of transcription 3
TAMs	Tumor associated macrophages
TCR	T-cell receptors
TGF	Transforming growth factor
TLR	Toll like receptors
TNF	Tumor necrosis factor
TRAMP	Transgenic adenocarcinoma mouse prostate
US	United States
VEGF	Vascular endothelial growth factor

3.1 Introduction

Inflammation is a physiological process that is initiated upon exposure to various infections or tissue injury. The inflammatory processes leads to a cascade of chemical events targeted towards eradication of pathogens, clearing tissue and cellular debris, regeneration of the epithelium and remodeling of the stroma. However, if this highly regulated process remains unchecked, or normal healthy tissue integrity is not restored and the inflammatory response persists, it results in significant cellular and genomic damage. The sustained inflammation generates a multitude of various reactive nitrogen and oxygen species, cytokines, chemokines, and growth factors. The persistent high level of these factors potentially leads to uncontrolled cellular proliferation and enhanced genomic instability. Genomic instability (e.g. activation of oncogenes and/or loss of tumor suppressors) coupled with unchecked cell proliferation due to presence of growth factors increases the risk of developing several types of malignancies, including prostate cancer (PCa) [1–3].

PCa is a leading public health concern that places a significant burden on healthcare systems worldwide [4]. PCa risk factors include family history, old age and ethnicity. Nearly 3 million men in the US currently live with the disease and approximately 14% of men will be diagnosed with PCa in their lifetime. This year alone in the US over 26,000 patients will die from PCa [4]. In Europe, there are approximately 346,000 new PCa cases and 87,000 deaths per year [5].

The longstanding observation and epidemiological link between inflammation and cancer has been recognized since the dawn of modern medicine when the German physician Rudolf Virchow in 1863 first described leucocyte infiltration and their distribution in neoplastic tissues. He further proposed that these "lymphoreticular infiltrates" in the tumor perhaps were remnants of chronic inflammation sites [6].

Given that less than 10% of all cancers are caused by germline mutations, there is great emphasis placed on understanding the underlying mechanisms that cause the vast majorities of human tumors; that is, those that initiate from acquired somatic mutations and detrimental environmental factors [7]. There exists a strong association between chronic inflammatory diseases such as gastritis, hepatitis, prostatitis or colitis and the increased risk of developing carcinomas in the afflicted organ. In fact, out of the nearly 600,000 thousand people who will die this year from hepatocellular carcinoma (HCC), 90% of the cases present with some form of hepatic injury and inflammation [8]. For men diagnosed with prostatitis around 18% of them will develop prostate cancer. Moreover, men who show signs of chronic inflammation in non-cancerous prostate tissue have nearly twice the risk of actually developing prostate cancer later on than those without inflammation [9]. This association is even stronger for those who ultimately develop high-grade disease (Gleason score $\geq$ 7) [10]. Overall approximately 20% of diagnosed adult cancers have been attributed to chronic inflammatory diseases [11].

3.1.1 Prostatic Inflammation and Cancer

A recent resurgence of interest into the tumor-promoting effects of the inflammatory microenvironment has been led by the abundance of clinical, molecular, histopathological and epidemiological-based evidence connecting prostate cancer and inflammation concurrence [12]. PCa development is mediated in part by hereditary components, but particularly in regards to inflammation-induced disease, also by environmental exposures such as infectious agents and dietary carcinogens. This is evidenced by the apparent increase in prostate cancer risk among men from geographic areas with low prostate cancer incidence (Southeast and East Asia) who immigrate to western countries [13]. While the exact sources of prostatic inflammation are still being investigated the environmental exposures that induce or increase the risk include (see also Fig. 3.1):

1) Infectious microorganisms such as *E. coli*, *Propionibacterium acnes* and others associated with the intraprostatic reflux of urine and sexually transmitted diseases (*Neisseria gonorrhoeae*) and prostatitis [14, 15]. Viruses such as Human papillomavirus (HPV), herpes simplex virus 2 (HSV2) and cytomegalovirus

Fig. 3.1 Various intrinsic and extrinsic factors can increase the risk of prostatic inflammation. See text for details

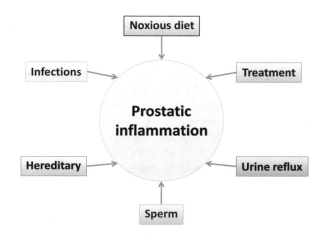

(CMV) can also infect the prostate [16, 17]. However, their frequency of infection and role in inflammation-induced carcinoma is largely unknown.

2) Noxious dietary elements and lifestyle-risk factors, including exposure to heterocyclic amines (HCAs) produced from cooking meat at high temperatures, estrogen, and obesity.

3) Treatment for reproductive ailments or tumor-induced inflammation [18].

4) Urine reflux causing chemical irritation with metabolites such as uric acid can lead to chronic inflammation within the prostate [19]. It can synergize with infection to further aggravate chronic inflammation.

5) Sperm seen in prostate tissue have been associated with PIA and inflammation. However, frequent ejaculation could potentially flush out urinary carcinogens and has been linked with decrease in prostate cancer incidence [20].

6) Hereditary inflammation-related genes such as macrophage inhibitory cytokine-1 (MIC1), Toll like receptors (TLRs), and interleukin receptor −1 antagonist (IL-1RN).

Inflammation Based Therapeutics Interestingly, studies have also found significant correlation between intake of anti-inflammatory compounds and reduced prostate cancer risk. In one study the anti-inflammatory phytoestrogens, genistein and daidzein, found in soy and green tea were associated with reduced risk of PCa [12], perhaps through modification of glutathione S-transferase P (GTSP1) and ephrin B2 (EphB2) promoter regions [12]. Furthermore, the use of aspirin and non-steroidal anti-inflammatory drugs (NSAIDs) has been associated with reduced risk of PCa, possibly through inhibition of the cyclooxygenase 2 (COX-2) enzyme [12, 21]. COX-2 is an inflammation related protein that facilitates the production of prostaglandins, which in turn promotes neoangiogenesis and cell migration, and reduces apoptosis [22]. Finally, the use of statins (prescribed primarily for lowering cholesterol) has been correlated with reduced risk of advanced and aggressive PCa by inhibiting 3-hydroxy-3-methyl-glutaryl-coenzyme A (HMG-CoA) [23, 24] (Fig. 3.1).

3.1.2   Proliferative Inflammatory Atrophy (PIA)

Proliferative inflammatory atrophy (PIA) are lesions within the prostate that are atrophic and are associated with increased acute or chronic inflammatory cell infiltration [25]. Some atrophic lesions show increased number of epithelial cells but are devoid of inflammatory cells. Morphological studies have shown association between the presence of these lesions and prostatic intraepithelial neoplasia (PIN) and carcinoma [25, 26]. Remarkably, these lesions are mostly found in the peripheral zone of the prostate where PCa most commonly occurs [27]. In fact, no cancerous lesions have been reported in the central zone of the prostate [28]. Although some evidence of molecular changes in PIA has been observed (such as GSTP1 hypermethylation), no studies have shown clonal genetic alterations in PIA [29]. Other genes such as NKX3.1 and CDKN1B, which are downregulated or lost in PIN and PCa, have also been shown to be downregulated in PIA [25, 30]. Indeed, targeted disruption of these genes in mouse models results in the development of PIN or invasive PCa [31].

3.1.3   Role of Innate and Adaptive Immunity

Inflammation is a process that involves the interplay between innate and adaptive immune responses following infection or injury and have a powerful influence on the development of the tumor microenvironment by producing a wide variety of pro-inflammatory oxidizing species, cytokines and chemokines, which results in carcinogenesis by promoting growth, angiogenesis, differentiation, survival and migration of tumor cells [32].

3.1.3.1   Immune Cells: Do they Help or Harm?

Although immune cells are important mediators of PCa progression, the incomplete phenotypic characterization of these cellular infiltrates combined with conflicting clinical evidence represents a gap of knowledge critical to understanding the intricate roles immune cells play in the tumor microenvironment [18]. One recent study analyzing lymphocyte infiltration in tumor tissues observed that both high and low levels of CD3+ cells (T-cell co-receptors) were correlated with reduced PSA recurrence-free survival [33]. Inclusion of other T-cell subtype markers (such as CD4+, CD8+, etc.) would have perhaps helped in parsing this discrepancy [18]. In normal prostates and prostates with benign prostatic hyperplasia (BPH), inflammatory cells have been shown to be comprised of CD3+ T cells (~70–80%) and B cells CD20+ (~15–20%), along with a high number of macrophages. Whereas normal prostates contain a greater fraction of CD8+ T-cells, inflammatory lesions are enriched in the CD4+ T cells subtype. Moreover, most of the observed T cells were alpha beta while less than 1% were gamma delta TCR positive T-cells. Additionally, 40% of the T cells present in inflammatory lesions were memory T-cells [34–36]. Regulatory T-cells (Tregs) are potent suppressors of anti-tumor adaptive immune responses, and higher amounts of CD4+, CD25+, and Foxp3+ Treg cells are found in the tumors and blood of prostate cancer patients with clinically localized disease [37]. With this in mind, a thorough analysis of various T-cell subsets in the context of various grades and stages of prostate cancer would shed immense new light on understanding the effects of immune cell presence in the tumor microenvironment.

Tumor-associated macrophages (TAMs) are another class of cells that are strongly associated with disease progression [38], yet only one study was able to determine the M2 macrophage subtype responsible for this phenomenon [39], which may explain the fact that despite the several studies documenting the pro-tumorigenic properties of TAMs only a handful of studies were able to directly correlate TAM infiltration with disease recurrence [33].

Future work aimed at completely characterizing the inflammatory cells within the tumor microenvironment will be essential in developing novel immunotherapies and identifying immune-based prognostic indicators.

3.1.3.2 Innate Immunity and Prostate Cancer

The innate immune system is the rapid-acting 'first line of defense' against pathogens and is comprised of several types of cells including granulocytes (i.e. neutrophils, basophils and eosinophils), dendritic cells, natural killer cells, macrophages, and mast cells. Mast cells and macrophages are probably the two most extensively studied innate immune cells in prostate cancer.

Mast cells are able to produce both pro- and anti-tumorigenic cytokines and as such are dynamic and effective regulators of the interactions within the tumor microenvironment. Their function is dependent on their environment, and probably varies in different types and stages of cancer [38]. For example some studies of PCa patients have noted high densities of intra-tumor infiltrated mast cells associating with overall lower Gleason-grade tissue and better prognosis [40, 41], yet in a separate study of PCa patients, low counts of mast cells were linked to lower Gleason scores and longer progression-free survival times [42]. These conflicting observations are better understood when considering that mast cells are capable of producing an immense number of distinct regulatory cytokines and effector molecules, including serotonin, heparin and proteases [43]. More evidence however is needed to support their roles in prostate carcinogenesis.

Macrophages represent another major class of immune cells that have been studied for their use in evaluating disease prognosis in PCa. Significantly greater levels of macrophage colony-stimulating factor (M-CSF) and colony-stimulating factor-1 receptor (CSF-1R) have been observed in tumor and stromal cells near primary tumors of patients with metastatic disease compared to those without metastases [44]. Macrophages may also promote prostate cancer invasion by secreting proteases (notably cathepsin K and cathepsin S) that breakdown the extracellular matrix.

In a study aimed at identifying proteins associated with tumor progression, cathepsin S was found to be upregulated in poorly-differentiated and metastatic tumors taken from TRAMP (transgenic adenocarcinoma mouse prostate) mice models of PCa as well as in high Gleason grade tumors from patients [45]. Another recent study demonstrated that tumor growth in bones was impaired in cathepsin K-deficient mice injected intra-tibially with PC3 cells (a PCa cell line model of castration-resistant disease) [46].

3.1.3.3 Adaptive Immunity and Prostate Cancer

Similarly to cells of the innate immune system, T and B lymphocytes of the adaptive immune system are known to have paradoxical roles in carcinogenesis, especially in chronically inflamed tumors and lesions that typically are associated with prolonged interactions with adaptive immune cells [38]. T-lymphocytes in particular have been frequently examined for their use as prognostic markers and in general have been associated with good prognosis as they may act to illicit an anti-tumor response [38]. The study of T-cells is hindered as they are normally differentiated by their cytokine secretions, which is difficult to analyze by immunohistochemistry, and more advanced approaches using flow cytometry to separate immune cells from prostate cells can be difficult to accomplish [38]. However, using flow cytometry for the phenotypic analysis of T-cells using serial needle aspirates of peripheral prostate tissue is certainly possible [47].

Although they are present in the tumor microenvironment [33], much less is known about the role of B-lymphocytes in prostate cancer. They have been reported to promote the progression of castration-resistant cancer cells by activating STAT3 and the proto-oncogene BMI1 [48].

3.2   Mechanisms of Inflammation Induced Carcinogenesis

Inflammation Genes in Prostate Cancer Several hereditary prostate cancer risk studies have revealed the potential involvement of inflammation-related genes as risk factors for hereditary prostate cancer. Linkage studies in families with prostate cancer have identified an E265X mutation in the innate immune response gene RNAse L (which is located on chromosome 1q and is involved in interferon (IFN) signaling) as a PCa-susceptibility gene in a family of European descent, and a M1I mutation in the same region in a family of African descent [49]. Another gene identified in families with PCa and associated with increased susceptibility is located on chromosome 8p and encodes a macrophage specific scavenger receptor (MSR1) [50]. Other inflammatory genes identified in Swedish case-control cohort studies which are associated with risk of developing PCa include MIC1, Toll like receptors (TLRs) and interleukin receptor −1 antagonist (IL-1RN) [51–53].

The inflammatory state can initiate or promote neoplastic progression if it is able to transform cells in the local environment into the full malignant phenotype. Phenotypic hallmarks include tissue remodeling, angiogenesis and metastasis. A tumor microenvironment rich in sustained proinflammatory cytokines and inflammatory cells can induce or promote neoplastic and malignant progression in several ways:

1) Generating reactive oxygen and nitrogen species that inflict cellular, epigenetic, and genetic alterations and damage
2) Sustaining the inflammatory tumor-microenvironment and recruiting additional leukocytes that promote angiogenesis, proliferation, vascular and tissue growth and remodeling
3) Elaborating the cytokine and chemokine network that promotes cell replication, differentiation and inhibition of apoptosis
4) Each mechanism has unique and significant contributions to carcinogenesis, and has been described in detail in the sections below.

3.2.1   Reactive Oxygen and Nitrogen Species (ROS and RNS) Generation

In response to infections, inflammatory cells (usually neutrophils and macrophages) synthesize a variety of toxic compounds designed to eradicate microorganisms. These compounds are reactive oxygen and reactive nitrogen species (ROS and RNS, respectively) and include hydrogen peroxide (H_2O_2), the hydroxyl radical (OH•), nitric oxide (NO), organic peroxides, singlet oxygen and the superoxide anion (O_2•-) [54, 55]. Under normal metabolic conditions the majority of the free radicals, ROS and RNS produced are byproducts of aerobic cellular respiration, generated in the intracellular milieu by mitochondria. However in response to pathogens, neutrophils and macrophages produce these compounds via extracellular membrane-bound enzyme complexes known as NADPH oxidases (i.e. NOX enzymes) and release ROS rapidly into the tumor microenvironment in an 'oxidative burst' [54]. During chronic inflammation there exists an imbalance between the amount of oxidizing agents produced and the host's ability to process them. The enzymes (e.g. superoxide dismutase, catalase, peroxiredoxin, thioredoxin glutathione reductase and glutathione S-transferase) and antioxidant molecules (e.g. glutathione, flavonoids and vitamins A, C and E) responsible for detoxifying the environment become overwhelmed, leading to a state of continued oxidative stress [56]. The buildup of these highly reactive compounds leads to significant mutagenic and genome-destabilizing DNA lesions; in fact there are over 100 oxidized DNA products currently known [57]. Some of these damages can either arrest or promote transcription, bring about point mutations, induce replication errors, and inhibit DNA repair [58]. One type of point mutation (a G to T transversion) has been observed in both Ras [59] and p53 genes [60] in multiple cancers, indicating that ROS and RNS may directly activate or inactivate proto-oncogenes and tumor suppressor genes, respectively.

Similar to mutations, epigenetic alterations can also contribute to carcinogenesis. The presence of reactive oxidizing species in the inflamed microenvironment has also been associated with epigenetic damage through aberrant DNA methylation and histone modifications [61]. Generally hypermethylation in the promoter regions of genes blocks transcription thereby regulating genetic expression within a cell. Exactly how ROS and RNS lead to increased DNA methylation is still unclear, however one proposed mechanism suggests that 5-halogenated cytosines formed from ROS can prevent DNA methyltransferases (i.e. DNMT1) from distinguishing methylated from halogenated cytosines leading to altered methylation [62]. Hypermethylation of putative tumor suppressor genes, such as RUNX3 in esophageal cancer [63] and GATA-4 and GATA-5 in colorectal and gastric cancers [64] leads to complete deactivation of their downstream targets. Moreover, DNA repair genes such as MLH1 and BRCA1 are also targets of hypermethylation, whereby their silencing results in accumulation of further genetic damage leading to the development of the malignant phenotype [65].

Lipids and proteins are also highly susceptible to damage from free radicals. Peroxidation of lipids generates lipid radicals and aldehydes, such as 4-hydroxy-2-nonenal (HNE), a well-characterized molecule known to affect the function of proteins involved in signaling pathways [66]. Oxidative damage to proteins can alter their structure and stability and commonly involves the carbonylation or nitrosylation of amino acid side chains. But the most severe and permanent damage results from disulfide bond–mediated protein cross-linkages or the formation of bulky protein aggregates [66]. Oxidizing species therefore pose a significant threat to the maintenance of structural integrity of both the cell membrane and many proteins involved in cell signaling and essential enzymatic pathways. Ultimately, the sum of these damages is important for the first step in carcinogenesis, the initiation stage, where normal cells acquire the right amount and type of mutations to help them survive and rapidly proliferate.

Deregulation of the transcription factor erythroid 2p45 (NF-E2)-related factor 2 (Nrf2) can also potentially contribute to ROS accumulation. Nrf2 is known to mediate the expression of several important antioxidant enzymes by interacting with the antioxidant-response element (ARE) promoter region of these genes. In fact, the expression of Nrf2 has been shown to be significantly downregulated in prostate tumors [67]. The loss of Nrf2 results in the suppression of glutathione-S-transferase (GST) expression (a target gene of Nrf2), leading to ROS accumulation and DNA damage in Nrf2-deficient cells [67]. GST itself is also prone to mutations, and its somatic silencing has been observed in nearly all cases of prostate cancer examined by Nelson and colleagues [68]. In fact, of all the genes known to be aberrantly methylated in PCa, GST is the most frequently methylated, with its methylation status positively correlating with both Gleason grade and tumor volume [69].

A significant portion of ROS come directly from the NOX family of enzymes [70]. Ectopic expression of the NOX1 isoform has been found to enhance the growth and tumorigenicity of prostate epithelial cells. Moreover, tumors that express NOX1 also overexpress VEGF and VEGF receptors, thereby vascularizing previously-dormant tumors and enabling their growth [71]. Conversely, downregulation of

NOX5 leads to dramatic growth inhibition and treatment with the NOX inhibitor diphenylene iodonium (DPI) caused cells to undergo apoptosis [72].

Age and Testosterone Age is a risk factor for PCa development (median age at diagnosis is 65) suggesting that changes in cellular metabolism might initiate the onset of PCa [70]. Advancing age has been associated with the increased risk of developing metabolic abnormalities that impairs a cell's ability to detoxify ROS leading to the development of PCa [70].

Steroid hormones (i.e. testosterone and dihydrotestosterone [DHT]) are critical for the proper maintenance and functioning of the prostate and have long been thought to regulate redox homeostasis within the tissue. Studies in rats have shown that castration induced the expression of NOX enzymes and reduced the expression of superoxide dismutase 2, glutathione peroxidase 1, thioredoxin, and peroxiredoxin 5 [73] Furthermore, replacement of testosterone levels decreased the NOX expression and restored the above-mentioned antioxidant enzymes to normal levels. Other studies have demonstrated that prostate cancer cells stimulated by androgens experience increased oxidative stress [74, 75]. While circulating levels of androgens can influence the production of ROS, exactly how their presence leads to redox imbalance is unclear.

3.2.2 The Inflammatory Tumor–Microenvironment

As mentioned previously, the innate immune system is the rapid-acting non-specific 'first line of defense' of the body and is comprised of cells that express on their surface Toll-like receptors (TLRs) that can recognize structurally conserved molecular domains found on microbes [76]. The activation of TLRs leads to a multitude of intracellular events including the activation NF-κB signaling pathways resulting in increased production of proinflammatory cytokines, chemokines and increased synthesis of nitric oxide (NO) [76]. The cytokines produced by innate immune cells alert the immune system to the presence of pathogens and promote the differentiation and activation of B and T lymphocytes, which are the major adaptive immune cells that are committed to the recognition of specific antigens [76].

Over time the accumulation of somatic mutations and other damage resulting from oxidative stress alters the growth and migration of epithelial cells. These epithelial cells, along with tumor cells, produce various cytokines and chemokines to attract leukocytes (i.e. dendritic cells, eosinophils, lymphocytes, macrophages, mast cells and neutrophils) to the affected area [77]. While these immune cells are all known to contribute to tumor angiogenesis, invasion, metastasis and proliferation [78], tumor-associated macrophages (TAMs) in particular have been associated with poor prognosis in several cancers and contribute to carcinogenesis in multiple ways [77, 79]. TAMs release interleukins and prostaglandins to suppress anti-tumor responses, and work to vascularize the tumor by releasing angiogenic factors such as endothelin-2 [80] and vascular endothelial growth factor (VEGF) [77]. TAMs can also stimulate tumor cell migration and proliferation by releasing several types of epidermal growth factors. Furthermore TAMs synthesize proteases such as cathepsins, matrix metalloproteinases (MMPs, i.e. MMP-2 and MMP-9) and urokinase-type plasminogen activator (uPA), which breakdown the basement membrane of cells and allows for the remodeling of the stromal matrix thereby promoting tumor cell invasion and metastasis [81]. Mast cells and neutrophils release many of the same, or similar, growth factors and proteases as macrophages and are therefore thought to contribute significantly to both angiogenesis and metastasis [77]. Moreover, tumor cells express vital pro-inflammatory transcription factors (e.g. STAT3, NFκB) [82]. These transcription factors in turn induce the production of key cytokines (IL-6 and TNF), chemokines (CCL-2 and CXCL12) and inflammatory enzymes (COX-2), thereby leading to a complex inflammatory microenvironment surrounding the tumor and infiltrated immune cells. Autocrine and paracrine cytokine signaling within the tumor microenvironment may lead to constitutive altered signaling leading to cancer-related inflammation which influences cell survival, proliferation, angiogenesis, invasion and metastasis, and immune suppressor phenotype [83]. Thus, the immune cells in the tumor microenvironment are able to directly transform the milieu into one that benefits the growth of tumor cells, by vascularizing and remodeling tumor tissues to firmly entrench them into the local environment, provide nutrients, and allow for the invasion of tumor cells into distant parts of the body (Fig. 3.3).

3.2.2.1 Inflammasomes

Inflammasomes are protein complexes assembled during heightened inflammation by cells of the innate immune system. A characteristic feature of inflammasomes is the presence of pattern recognition receptors (PRRs) [such as Toll-like receptors (TLRs) and nucleotide-binding oligomerization domain (NOD) receptors (NLRs)] which recognize pathogen-associated molecular patterns (PAMP) or danger-associated molecular patterns (DAMP) [84]. These complexes regulate caspase-1, which promotes an inflammatory response through activation and secretion of IL-1β and IL18 and induction of pyroptosis, an immune-regulated form of programmed cell death [85, 86]. Once activated by caspase-1, IL-18 induces IFN-γ production in NK cells and T-cells. This, in turn, enables anti-pathogen responses by macrophages including the production of reactive oxygen and nitrogen species [87, 88]. Unregulated expression of IL-1 has been indicated in malignancies and CD4+ T-cell production [89, 90]. CD4+ T cells produce IL-17, which in conjunction with IL-23, progresses skin carcinogenesis [90]. On the other hand, lack of IL-1 signaling has been shown to inhibit tumorigenesis by increasing myeloid-derived suppressor cell (MDSC) infiltration [91]. Also, IL-18-mediated IFN-γ production can also limit carcinogenesis, as seen in murine colorectal cancer [88, 92]. Finally, radiotherapy and chemotherapy can promote inflammasome activity, which in turn may stimulate antitumor immune responses [93]. Since inflammasomes play a role in both the protection and progression of cancer further research is needed to parse out the roles of various members of the complex and the downstream pathways affected in order to harness their protective capacities for therapeutic purposes.

3.2.2.2   microRNAs and Inflammation

MicroRNAs are non-coding RNA molecules usually 19–24 nucleotides long that form hairpin-like structrues and play an instrumental role in post-transcriptional regulation, either through the degradation of mRNA or by blocking translation thereby affecting cellular processes such as cell growth, angiogenesis, immune response and survival [94–96]. Inflammation mediated production of reactive oxygen species can cause genomic instability and production of aberrant miRNAs [97]. Additionally, NF-κB and other transcription factors can regulate the expression of genes that code for miRNAs [98]. Aberrant regulation of certain miRNAs can cause an increase in oncogene expression, or suppression of tumor suppressors or both, Fig. 3.2. IL-6 produced by immune cells stimulates miR-21 (a microRNA highly expressed in inflammatory diseases and responsible for carcinogenesis) through NF-κB [99]. Incidentally, inhibition of miR-21 results in tumor regression in xenograft mouse models [100]. In breast cancer, Let-7 miRNAs regulate the level of IL6 and its inhibition through NF-κB constitutes a positive feedback loop ultimately resulting in further IL6 production [101]. miR-155, another oncogene, is highly expressed in inflammatory conditions such as *H. pylori* and EBV infections and inhibits TP53-induced nuclear protein 1 (TP53INP1), a pro-apoptotic gene, leading to increased tumor cell survival [102, 103]. On the other hand, miR-663 behaves as a tumor-suppressor and its loss in gastric cancer increases oncogenesis [104]. miR-146 also shows tumor suppressor properties and its overexpression results in reduction in the levels of IL6 and IL8 [105]. In prostate cancer, miR-146 has been identified to have tumor suppressor properties through its inhibitory effect on Rac1 [106]. Also, overexpression of miR-101 results in the inhibition of prostate cancer cell growth [107]. In metastatic prostate cancer, loss of miR-101 results in up-regulation of EZH2, an E-cadherin silencer [108]. Since miRNAs exhibit distinct expression patterns in drug-resistant cancers, they may be used to differentiate between drug-sensitive and insensitive malignancies [109, 110]. Although microRNAs constitute a small fraction of the genome, growing evidence suggests that they play a substantial role in inflammation related-cancers and may hold diagnostic, prognostic, and therapeutic value.

3.2.3   The Network of Cytokines and Chemokines (Fig. 3.2)

The complex system of chemokine and cytokine signaling between stromal, tumor, and immune cells are involved in promoting cell replication and survival within the inflammatory environment. Cytokines can be classified as proteins, peptides, or glycoproteins that are secreted or are membrane-bound and regulate the differentiation

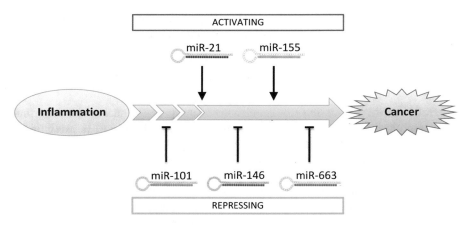

Fig. 3.2 Simplified schematic showing that miRNAs can be 'oncogenic' or 'tumor suppressive', based on whether they are lost (resulting in increased expression of oncogenes) or are overexpressed (resulting in the suppression of tumor suppressors), respectively

and activation of immune cells. These include interleukins, growth factors, TNF-α and colony-stimulating factors [111].

Several ILs are known to associate with the diseased prostate such as IL-1, IL-4, IL-5, IL-6, IL-8, IL-10, IL-13, IL-17, IL-23, TGF-β and TNF-α [112, 113]. While some cytokines (i.e. IL-1 and IL-4) are primarily associated with the development of benign prostatic hyperplasia (BPH) [114, 115] others such as IL-6, IL-8, TGF-β and TNF-α are known to directly contribute to carcinogenesis. Perhaps one of the best-studied pro-inflammatory cytokines in cancer, IL-6 was first discovered to enhance the proliferation of intestinal epithelial cells and was elevated in the serum of colon cancer patients [77]. Patients with PCa display high levels of IL-6 and its soluble receptor in the circulating plasma [116]. A crosstalk between IL-6 and androgen receptor activation has also been observed [117]. Remarkably, an androgen-sensitive PCa cell line (LNCaP) that was continuously exposed to IL-6 in vitro developed neuroendocrine features [118] and has been thought to be a factor driving the neuroendocrine phenotype in prostate tumors [119].

Incidentally mutations in Ras and TP53 also lead to increased production of IL-6 [120, 121]. IL-6 is also a potent activator of members of the Janus kinase (JAK) family of tyrosine kinases, which in turn further activate transcription factors known as signal transducers and activators of transcription (STATs), especially STAT3 [122]. STAT3 is constitutively active in many cancers and promotes cell proliferation by upregulating the expression of cyclins, the proto-oncogene c-Myc, and anti-apoptotic genes such as Bcl-2, Bcl-X_L and survivin [113]. Besides activating the STAT3 pathway, depending on the cellular context IL-6 can also signal through the mitogen-activated protein kinase (MAPK) and phosphatidylinositol-3 kinase (PI3K) compensatory signaling pathways that are upregulated in castration-resistant PCa cell lines [123].

From a therapeutic standpoint, Siltuximab, a chimeric humanized antibody, has been shown to have high specificity and affinity for binding to IL-6 [124]. In in vitro studies, the antibody sensitized PCa cancer cell lines to cis-diamminedichloroplatinum- and etoposide- mediated cytotoxicity [125]. This antibody has been shown to be safe for combination therapy with docetaxel in a Phase I study of mCRPC patients where 62% patients showed reduction in serum PSA [126].

Tumor necrosis factor (TNF) is another major cytokine involved in systemic inflammation, especially during the early events of carcinogenesis by recruiting inflammatory cytokines, growth factors, and epithelial adhesion factors to damaged tissue [77]. The process of angiogenesis is supported by TNF via the induction of various angiogenic factors (e.g. VEGF, basic fibroblast growth factor) and enzymes (thymidine phosphorylase) [127]. TNF is also a major inducer of nuclear factor-κB (NF-κB), a transcription factor that upregulates many of the same pro-replication, pro-survival genes as STAT3. NF-κB is a dimer formed by Rel family proteins (i.e. RelA/p65, RelB, c-Rel, NF-κB1/p50, and NF-κB2/p52) and is held in an inactive conformation in the cytoplasm by inhibitory IκB proteins. Upon activation by external stimuli, such as TNF, the dimer is released and it enters the nucleus where it binds to the promoter of NF-κB-responsive genes [128, 129]. In PCa, the NF-κB pathway is dysregulated resulting in the progression to the androgen-independent state that ultimately leads to lethal CRPC. Constitutive NF-κB activation has been reported in prostate tumors [130] and the active, nuclear-localized NF-κB has been observed in organ-confined prostate tumors, but not in benign tissues, suggesting that constitutive NF-κB activation may also be an important early event in prostate carcinogenesis [131] (Fig. 3.3).

3.2.3.1 EMT-Linking Inflammation and Cancer

Growing evidence suggests that the tumor microenvironment transmits inflammatory signals that enhance the metastatic capacity of cancer through the activation of a developmental process known as epithelial to mesenchymal transition (EMT). As mentioned previously, immune cells including DCs, TAMs, NK cells, regulatory T cells, neutrophils, B cells and MDSCs constitute a considerable proportion of the tumor microenvironment and behave as mediators of inflammation-induced EMT. For EMT to occur, cells must have the capacity to undergo this process irrespective of their oncogenic content and have sufficient signals that promote EMT induction [132]. Moreover studies have demonstrated that epithelial cells can undergo various degrees of EMT when induced with TGF-β1 resulting in the activation of SMAD transcription factors that then stimulates EMT proteins such as Snail, Zeb, and AP-1 [133, 134]. Additionally, Wnt signaling sensitizes cells to TGF-β induced EMT by inhibiting GSK3-β, which inactivates Snail, Zeb, and β-catenin via phosphorylation [135]. TGF-β also interacts with BMP7, a promoter of epithelial cell differentiation that impedes metastasis in prostate and breast cancer [136]. Additionally, oncogenic pathways such as Ras, Notch, and Hedgehog,

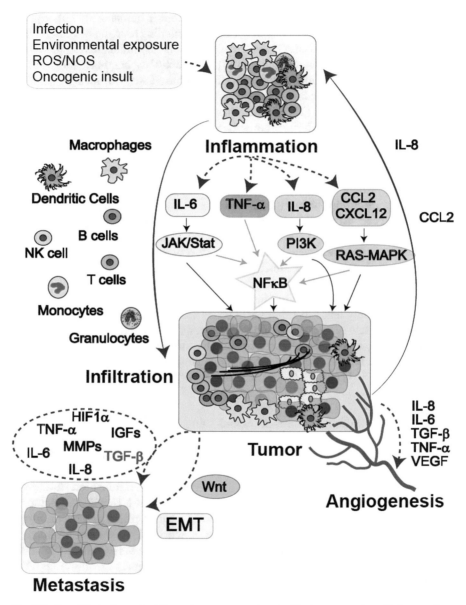

Fig. 3.3 Simplified schematic of the various processes involved in the progression of inflammation-initiated tumorigenesis

have been shown to be involved in stemness through the induction of EMT by TGF-β [137–140].

Within the tumor microenvironment, TAMs, MDSCs, and Tregs produce TGF-β1, exacerbating cancer to a more aggressive and invasive state [141, 142].

Reciprocally, tumor cells produce TGF-β1 for immune cell recruitment and polarization; these cells collectively form a tumor-permissive microenvironment that drives EMT [132]. In prostate cancer, IGF-1 induction by TGF-β1 has been linked to EMT [143]. The NF-κB pathway, a Snail-1 stabilizer, is activated by TNF-α produced by TAMs and enhances TGF-β induced EMT [144, 145]. Although inflammation induced EMT is a well-established phenomenon; little is known about the migratory behavior of disseminated tumor cells. In fact, tumor cells undergoing EMT may co-migrate with macrophages by acquiring immune cell properties, mainly chemotaxis [146].

3.2.3.2 Stem Cell Theory

Tumors are composed of an array of cell types with differential tumorigenic capacities. Cancer stem cells (CSCs) are a subpopulation of cells within the tumor that have been implicated as key drivers of carcinogenesis. The defining characteristics of CSCs are their ability to differentiate into cells that reinforce the oncogenic phenotypes of the tumor, as well as their ability to self-renew and sustain tumor growth [147, 148]. Since inflammatory signals have been shown to induce processes that regulate normal stem cells, they may also play a critical role in the initiation and maintenance of CSCs [149]. NFκ-B signaling, activated by TNF-a, has been indicated in neural stem cell proliferation and inhibition of differentiation [150]. IL-6 can supplement the self-renewal of hematopoietic stem cells [151]. Additionally, low levels of oxygen and ROS enhance the self-renewal capacity of normal stem cells. Deregulation of pathways that maintain levels of oxygen and ROS can impair stem-cell function [152–157]. Inflammatory cytokines produce high levels of ROS, and interestingly, cancer stem cells in acute myeloid leukemia (AML) have higher level of ROS compared to normal stem cells [158]. It is possible that high ROS levels may lead to deregulated stem-cell mechanisms—perhaps through genomic instability—that give rise to cancer stem cells. This discrepancy between normal stem cells and CSCs may hold significant therapeutic value [159–161]. Several studies have suggested a strong relationship between CSCs and EMT through the well-defined TGF-β pathway [162]. Sustained expression of Snail by TGF-β treatment on breast cancer cell lines leads to loss of E-cadherin and a phenotype synonymous to breast CSCs [163–165]. HIF-1α activation in hypoxic conditions—a CSC-friendly environment—can promote EMT through Twist-1 expression [166, 167]. Furthermore, developmental pathways such as Wnt and Hedgehog (Hh) have been linked to self-renewal and CSCs [168, 169]. Thus, inflammation may enhance tumorigenic potential and drive disease progression through the initiation, and subsequent utilization, of stem cell properties. Discovery of CSC-specific markers may supplement current diagnostic and prognostic tools for disease detection, monitoring and potentially be of therapeutic value.

3.3   Conclusions

There exists an association between the incidence of chronic inflammation and the ability of the inflammatory microenvironment to initiate or promote prostate carcinogenesis. However, the molecular details surrounding inflammation and prostate cancer are still far from being completely understood. This is particularly important in the case of the immune cells surrounding the tumor. There exists a fine balance between their protective versus aggravating role, and more work is needed to specifically identify the subpopulations of the various immune cells that contribute to each of the scenarios. Certainly, elucidation of the molecular and immunobiological mechanisms linking inflammation and PCa will be beneficial to the development of novel therapies and prognostic markers to treat and detect inflammation-associated malignancies of the prostate.

Acknowledgements   We would like to acknowledge Victoria Hackert for her help in collecting relevant materials. Both KKY and SSY are Prostate Cancer Foundation Young Investigators. AKT is supported by grants from PCF and Deane Prostate Health.

References

1. Albini A, Sporn MB (2007) The tumour microenvironment as a target for chemoprevention. Nat Rev Cancer **7**(2)
2. Gonzalgo ML, Isaacs WB (2003) Molecular pathways to prostate cancer. J Urol 170(6 Pt 1):2444–2452
3. Pihan GA et al (2003) Centrosome abnormalities and chromosome instability occur together in pre-invasive carcinomas. Cancer Res 63(6):1398–1404
4. Savage L (2007) Unreported VA data may affect SEER research, Cancer surveillance, and statistics gathering. JNCI J. Natl. Cancer Inst 99(23):1744–1752
5. Ferlay J et al (2006) Estimates of the cancer incidence and mortality in Europe in 2006. Ann Oncol 18(3):581–592
6. Virchow R (1881) An address on the value of pathological experiments. Br Med J 2(1075):198–203
7. Elinav E et al (2013) Inflammation-induced cancer: Crosstalk between tumours, immune cells and microorganisms. Nat Rev Cancer 13(11):759–771
8. Bishayee A (2014) The Inflammation and Liver Cancer, in Advances in Experimental Medicine and Biology. Springer Nature, pp 401–435
9. Sfanos KS, Isaacs WB, De Marzo AM (2013) Infections and inflammation in prostate cancer. Am J Clin Exp Urol 1(1):3–11
10. Gurel B et al (2014) Chronic inflammation in benign prostate tissue is associated with high-grade prostate Cancer in the placebo arm of the prostate Cancer prevention trial. Cancer Epidemiol Biomark Prev 23(5):847–856
11. Mantovani A et al (2008) Cancer-related inflammation. Nature 454(7203):436–444
12. Sfanos KS, De Marzo AM (2011) Prostate cancer and inflammation: the evidence. Histopathology 60(1):199–215
13. Lee J et al (2007) Cancer incidence among Korean-American immigrants in the United States and native Koreans in South Korea. Cancer Control 14(1):78–85

14. Pelouze PS (1935) Obscure pseudomembranous Trigonitis: Trigonitis Areata Alba. Ann Surg 101(1):594–598
15. Bushman W (2000) In: Lepor H (ed) Etiology of prostate. Prostatic diseases. W B Saunders Company, Philadelphia, pp 550–557
16. Zambrano A et al (2002) Detection of human polyomaviruses and papillomaviruses in prostatic tissue reveals the prostate as a habitat for multiple viral infections. Prostate 53(4):263–276
17. Samanta M et al (2003) High prevalence of human cytomegalovirus in prostatic intraepithelial neoplasia and prostatic carcinoma. J Urol 170(3):998–1002
18. Strasner A, Karin M (2015) Immune infiltration and prostate Cancer. Front Oncol 5
19. Persson BE, Ronquist G (1996) Evidence for a mechanistic association between nonbacterial prostatitis and levels of urate and creatinine in expressed prostatic secretion. J Urol 155(3):958–960
20. Leitzmann MF et al (2004) Ejaculation frequency and subsequent risk of prostate cancer. JAMA 291(13):1578–1586
21. Vidal AC et al (2014) Aspirin, NSAIDs, and risk of prostate cancer: Results from the REDUCE study. Clin Cancer Res 21(4):756–762
22. Masferrer JL et al (2000) Antiangiogenic and antitumor activities of cyclooxygenase-2 inhibitors. Cancer Res 60(5):1306–1311
23. Boudreau DM, Yu O, Johnson J (2010) Statin use and cancer risk: a comprehensive review. Expert Opin Drug Saf 9(4):603–621
24. Blake GJ, Ridker PM (2000) Are statins anti-inflammatory? Curr Control Trials Cardiovasc Med 1(3):161–165
25. De Marzo AM et al (1999) Proliferative inflammatory atrophy of the prostate: implications for prostatic carcinogenesis. Am J Pathol 155(6):1985–1992
26. Montironi R, Mazzucchelli R, Scarpelli M (2002) Precancerous lesions and conditions of the prostate: from morphological and biological characterization to chemoprevention. Ann N Y Acad Sci 963:169–184
27. Rich AR (1979) Classics in oncology. On the frequency of occurrence of occult carcinoma of the prostate: Arnold rice Rich, M.D., Journal of urology 33:3, 1935. CA Cancer J Clin 29(2):115–119
28. McNeal JE et al (1988) Zonal distribution of prostatic adenocarcinoma. Correlation with histologic pattern and direction of spread. Am J Surg Pathol 12(12):897–906
29. Nakayama M et al (2003) Hypermethylation of the human glutathione S-transferase-pi gene (GSTP1) CpG island is present in a subset of proliferative inflammatory atrophy lesions but not in normal or hyperplastic epithelium of the prostate: a detailed study using laser-capture microdissection. Am J Pathol 163(3):923–933
30. Bethel CR et al (2006) Decreased NKX3.1 protein expression in focal prostatic atrophy, prostatic intraepithelial neoplasia, and adenocarcinoma: association with Gleason score and chromosome 8p deletion. Cancer Res 66(22):10683–10690
31. Abate-Shen C, Shen MM (2002) Mouse models of prostate carcinogenesis. Trends Genet 18(5):S1–S5
32. Coussens LM, Werb Z (2002) Inflammation and cancer. Nature 420(6917):860–867
33. Flammiger A et al (2012) Intratumoral T but not B lymphocytes are related to clinical outcome in prostate cancer. APMIS 120(11):901–908
34. Steiner GE et al (2002) The picture of the prostatic lymphokine network is becoming increasingly complex. Rev Urol 4(4):171–177
35. Steiner GE et al (2003) Expression and function of pro-inflammatory interleukin IL-17 and IL-17 receptor in normal, benign hyperplastic, and malignant prostate. Prostate 56(3):171–182
36. Steiner GE et al (2003) Cytokine expression pattern in benign prostatic hyperplasia infiltrating T cells and impact of lymphocytic infiltration on cytokine mRNA profile in prostatic tissue. Lab Investig 83(8):1131–1146

37. Miller AM et al (2006) CD4+CD25high T cells are enriched in the tumor and peripheral blood of prostate cancer patients. J Immunol 177(10):7398–7405
38. Sfanos KS, Hempel HA, De Marzo AM (2014) The role of inflammation in prostate cancer, in advances in experimental medicine and biology. Springer Nature, pp 153–181
39. Lanciotti M et al (2014) The role of M1 and M2 macrophages in prostate Cancer in relation to extracapsular tumor extension and biochemical recurrence after radical prostatectomy. Biomed Res Int 2014:1–6
40. Johansson A et al (2010) Mast cells are novel independent prognostic markers in prostate Cancer and represent a target for therapy. Am J Pathol 177(2):1031–1041
41. Fleischmann A et al (2009) Immunological microenvironment in prostate cancer: high mast cell densities are associated with favorable tumor characteristics and good prognosis. Prostate 69(9):976–981
42. Nonomura N et al (2007) Decreased number of mast cells infiltrating into needle biopsy specimens leads to a better prognosis of prostate cancer. Br J Cancer 97:952–956
43. Khazaie K et al (2011) The significant role of mast cells in cancer. Cancer Metastasis Rev 30(1):45–60
44. Richardsen E et al (2008) The prognostic impact of M-CSF, CSF-1 receptor, CD68 and CD3 in prostatic carcinoma. Histopathology 53(1):30–38
45. Lindahl C et al (2009) Increased levels of macrophage-secreted cathepsin S during prostate cancer progression in TRAMP mice and patients. Cancer Genomics Proteomics 6(3):149–159
46. Herroon MK et al (2012) Macrophage cathepsin K promotes prostate tumor progression in bone. Oncogene 32(12):1580–1593
47. Sfanos KS et al (2008) Phenotypic analysis of prostate-infiltrating lymphocytes reveals TH17 and Treg skewing. Clin Cancer Res 14(11):3254–3261
48. Ammirante M et al (2010) B-cell-derived lymphotoxin promotes castration-resistant prostate cancer. Nature 464(7286):302–305
49. Smith JR et al (1996) Major susceptibility locus for prostate cancer on chromosome 1 suggested by a genome-wide search. Science 274(5291):1371–1374
50. Xu J et al (2002) Germline mutations and sequence variants of the macrophage scavenger receptor 1 gene are associated with prostate cancer risk. Nat Genet 32(2):321–325
51. Zheng SL et al (2004) Sequence variants of toll-like receptor 4 are associated with prostate cancer risk: results from the CAncer prostate in Sweden study. Cancer Res 64(8):2918–2922
52. Lindmark F et al (2004) H6D polymorphism in macrophage-inhibitory cytokine-1 gene associated with prostate cancer. J Natl Cancer Inst 96(16):1248–1254
53. Lindmark F et al (2005) Interleukin-1 receptor antagonist haplotype associated with prostate cancer risk. Br J Cancer 93(4):493–497
54. Khanna RD et al (2014) Inflammation, free radical damage, oxidative stress and cancer. Microinflammation 1:109. https://doi.org/10.4172/2381-8727.1000109
55. Klaunig JE, Kamendulis LM, Hocevar BA (2009) Oxidative stress and oxidative damage in carcinogenesis. Toxicol Pathol 38(1):96–109
56. Liou G-Y, Storz P (2010) Reactive oxygen species in cancer. Free Radic Res 44(5):479–496
57. Federico A et al (2007) Chronic inflammation and oxidative stress in human carcinogenesis. Int J Cancer 121(11):2381–2386
58. Maynard S et al (2008) Base excision repair of oxidative DNA damage and association with cancer and aging. Carcinogenesis 30(1):2–10
59. Bos JL (1988) The ras gene family and human carcinogenesis. Mutation Research/Reviews in Genetic Toxicology 195(3):255–271
60. Takahashi T et al (1989) p53: a frequent target for genetic abnormalities in lung cancer. Science 246(4929):491–494
61. Franco R et al (2008) Oxidative stress, DNA methylation and carcinogenesis. Cancer Lett 266(1):6–11
62. Valinluck V, Sowers LC (2007) Endogenous cytosine damage products Alter the site selectivity of human DNA maintenance methyltransferase DNMT1. Cancer Res 67(3):946–950

63. Long C et al (2007) Promoter Hypermethylation of the RUNX3 gene in esophageal squamous cell carcinoma. Cancer Investig 25(8):685–690
64. Akiyama Y et al (2003) GATA-4 and GATA-5 transcription factor genes and potential downstream antitumor target genes are epigenetically silenced in colorectal and gastric Cancer. Mol Cell Biol 23(23):8429–8439
65. Sharma S, Kelly TK, Jones PA (2009) Epigenetics in cancer. Carcinogenesis 31(1):27–36
66. Trachootham D et al (2008) Redox regulation of cell survival. Antioxid Redox Signal 10(8):1343–1374
67. Frohlich DA et al (2008) The role of Nrf2 in increased reactive oxygen species and DNA damage in prostate tumorigenesis. Oncogene 27(31):4353–4362
68. Nelson WG et al (2004) The role of inflammation in the pathogenesis of prostate cancer. J Urol 172(5):S6–S12
69. Zhou M et al (2004) Quantitative GSTP1 methylation levels correlate with Gleason grade and tumor volume in prostate needle biopsies. J Urol 171(6):2195–2198
70. Khandrika L et al (2009) Oxidative stress in prostate cancer. Cancer Lett 282(2):125–136
71. Arbiser JL et al (2002) Reactive oxygen generated by Nox1 triggers the angiogenic switch. Proc Natl Acad Sci 99(2):715–720
72. Brar SS et al (2003) NOX5 NAD(P)H oxidase regulates growth and apoptosis in DU 145 prostate cancer cells. AJP: Cell Physiol 285(2):C353–C369
73. Tam NNC et al (2003) Androgenic regulation of oxidative stress in the rat prostate. Am J Pathol 163(6):2513–2522
74. Pathak S et al (2008) Androgen manipulation alters oxidative DNA adduct levels in androgen-sensitive prostate cancer cells grown in vitro and in vivo. Cancer Lett 261(1):74–83
75. Miyake H et al (2004) Oxidative DNA damage in patients with prostate cancer and its response to treatment. J Urol 171(4):1533–1536
76. de Visser KE, Coussens LM (2005) The interplay between innate and adaptive immunity regulates cancer development. Cancer Immunol Immunother 54(11):1143–1152
77. Lu H (2006) Inflammation, a key event in cancer development. Mol Cancer Res 4(4):221–233
78. Coussens LM, Werb Z (2001) Inflammatory cells and Cancer. J Exp Med 193(6):F23–F26
79. Lin EY, Pollard JW (2004) Role of infiltrated leucocytes in tumour growth and spread. Br J Cancer 90(11):2053–2058
80. Grimshaw MJ, Wilson JL, Balkwill FR (2002) Endothelin-2 is a macrophage chemoattractant: implications for macrophage distribution in tumors. Eur J Immunol 32(9):2393–2400
81. Pollard JW (2004) Opinion: tumour-educated macrophages promote tumour progression and metastasis. Nat Rev Cancer 4(1):71–78
82. Grivennikov SI, Karin M (2010) Inflammation and oncogenesis: a vicious connection. Curr Opin Genet Dev 20(1):65–71
83. Crusz SM, Balkwill FR (2015) Inflammation and cancer: advances and new agents. Nat Rev Clin Oncol 12(10):584–596
84. Medzhitov R (2009) Approaching the asymptote: 20 years later. Immunity 30(6):766–775
85. Martinon F, Burns K, Tschopp J (2002) The inflammasome: a molecular platform triggering activation of inflammatory caspases and processing of proIL-beta. Mol Cell 10(2):417–426
86. Chen KW et al (2014) In: Hiraku YU, Kawanishi SO, Oshima H (eds) Inflammasomes and inflammation, in cancer and inflammation mechanisms : chemical, biological, and clinical aspects, p 1 online resource (409 pages)
87. Gu Y et al (1997) Activation of interferon-gamma inducing factor mediated by interleukin-1beta converting enzyme. Science 275(5297):206–209
88. Schroder K et al (2004) Interferon-gamma: an overview of signals, mechanisms and functions. J Leukoc Biol 75(2):163–189
89. Scheede-Bergdahl C et al (2012) Is IL-6 the best pro-inflammatory biomarker of clinical outcomes of cancer cachexia? Clin Nutr 31(1):85–88
90. Langowski JL et al (2006) IL-23 promotes tumour incidence and growth. Nature 442(7101):461–465

91. Bunt SK et al (2007) Reduced inflammation in the tumor microenvironment delays the accumulation of myeloid-derived suppressor cells and limits tumor progression. Cancer Res 67(20):10019–10026
92. Zaki MH et al (2010) IL-18 production downstream of the Nlrp3 inflammasome confers protection against colorectal tumor formation. J Immunol 185(8):4912–4920
93. Ghiringhelli F et al (2009) Activation of the NLRP3 inflammasome in dendritic cells induces IL-1beta-dependent adaptive immunity against tumors. Nat Med 15(10):1170–1178
94. Gong Z et al (2014) In: Hiraku YU, Kawanishi SO, Oshima H (eds) MicroRNAs and Inflammation-related cancer, in cancer and inflammation mechanisms : chemical, biological, and clinical aspects, p 1 online resource (409 pages)
95. Sonkoly E, Pivarcsi A (2009) microRNAs in inflammation. Int Rev Immunol 28(6):535–561
96. Schetter AJ, Heegaard NH, Harris CC (2010) Inflammation and cancer: interweaving microRNA, free radical, cytokine and p53 pathways. Carcinogenesis 31(1):37–49
97. Karin M, Greten FR (2005) NF-kappaB: linking inflammation and immunity to cancer development and progression. Nat Rev Immunol 5(10):749–759
98. Murdoch C et al (2008) The role of myeloid cells in the promotion of tumour angiogenesis. Nat Rev Cancer 8(8):618–631
99. Loffler D et al (2007) Interleukin-6 dependent survival of multiple myeloma cells involves the Stat3-mediated induction of microRNA-21 through a highly conserved enhancer. Blood 110(4):1330–1333
100. Si ML et al (2007) miR-21-mediated tumor growth. Oncogene 26(19):2799–2803
101. Iliopoulos D, Hirsch HA, Struhl K (2009) An epigenetic switch involving NF-kappaB, Lin28, Let-7 MicroRNA, and IL6 links inflammation to cell transformation. Cell 139(4):693–706
102. Motsch N et al (2007) Epstein-Barr virus-encoded latent membrane protein 1 (LMP1) induces the expression of the cellular microRNA miR-146a. RNA Biol 4(3):131–137
103. Gironella M et al (2007) Tumor protein 53-induced nuclear protein 1 expression is repressed by miR-155, and its restoration inhibits pancreatic tumor development. Proc Natl Acad Sci U S A 104(41):16170–16175
104. Pan J et al (2010) Tumor-suppressive mir-663 gene induces mitotic catastrophe growth arrest in human gastric cancer cells. Oncol Rep 24(1):105–112
105. Bhaumik D et al (2009) MicroRNAs miR-146a/b negatively modulate the senescence-associated inflammatory mediators IL-6 and IL-8. Aging (Albany NY) 1(4):402–411
106. Sun Q et al (2014) miR-146a functions as a tumor suppressor in prostate cancer by targeting Rac1. Prostate 74(16):1613–1621
107. Hao Y et al (2011) Enforced expression of miR-101 inhibits prostate cancer cell growth by modulating the COX-2 pathway in vivo. Cancer Prev Res (Phila) 4(7):1073–1083
108. Varambally S et al (2008) Genomic loss of microRNA-101 leads to overexpression of histone methyltransferase EZH2 in cancer. Science 322(5908):1695–1699
109. Blower PE et al (2008) MicroRNAs modulate the chemosensitivity of tumor cells. Mol Cancer Ther 7(1):1–9
110. Zheng T et al (2010) Role of microRNA in anticancer drug resistance. Int J Cancer 126(1):2–10
111. Dranoff G (2004) Cytokines in cancer pathogenesis and cancer therapy. Nat Rev Cancer 4(1):11–22
112. Vasto S et al (2012) Inflammation and cancer of the prostate, in prostate cancer: a comprehensive perspective. Springer Nature, pp 115–122
113. Grivennikov SI, Greten FR, Karin M (2010) Immunity, inflammation, and Cancer. Cell 140(6):883–899
114. Giri D, Ittmann M (2000) Interleukin-1α is a paracrine inducer of FGF7, a key epithelial growth factor in benign prostatic hyperplasia. Am J Pathol 157(1):249–255
115. Furbert-Harris P et al (2003) Inhibition of prostate cancer cell growth by activated eosinophils. Prostate 57(2):165–175

116. Nguyen DP, Li J, Tewari AK (2014) Inflammation and prostate cancer: the role of interleukin 6 (IL-6). BJU Int 113(6):986–992
117. Culig Z, Bartsch G, Hobisch A (2002) Interleukin-6 regulates androgen receptor activity and prostate cancer cell growth. Mol Cell Endocrinol 197(1–2):231–238
118. Deeble PD et al (2001) Interleukin-6- and cyclic AMP-mediated signaling potentiates neuro-endocrine differentiation of LNCaP prostate tumor cells. Mol Cell Biol 21(24):8471–8482
119. Sottnik JL et al (2011) The PCa tumor microenvironment. Cancer Microenviron 4(3):283–297
120. Ancrile B, Lim KH, Counter CM (2007) Oncogenic Ras-induced secretion of IL6 is required for tumorigenesis. Genes Dev 21(14):1714–1719
121. Yonish-Rouach E et al (1991) Wild-type p53 induces apoptosis of myeloid leukaemic cells that is inhibited by interleukin-6. Nature 352(6333):345–347
122. Hodge DR, Hurt EM, Farrar WL (2005) The role of IL-6 and STAT3 in inflammation and cancer. Eur J Cancer 41(16):2502–2512
123. Culig Z et al (2005) Interleukin-6 regulation of prostate cancer cell growth. J Cell Biochem 95(3):497–505
124. van Zaanen HC et al (1998) Chimaeric anti-interleukin 6 monoclonal antibodies in the treatment of advanced multiple myeloma: a phase I dose-escalating study. Br J Haematol 102(3):783–790
125. Borsellino N, Belldegrun A, Bonavida B (1995) Endogenous interleukin 6 is a resistance factor for cis-diamminedichloroplatinum and etoposide-mediated cytotoxicity of human prostate carcinoma cell lines. Cancer Res 55(20):4633–4639
126. Hudes G et al (2013) A phase 1 study of a chimeric monoclonal antibody against interleukin-6, siltuximab, combined with docetaxel in patients with metastatic castration-resistant prostate cancer. Investig New Drugs 31(3):669–676
127. Balkwill F (2002) Tumor necrosis factor or tumor promoting factor? Cytokine Growth Factor Rev 13(2):135–141
128. Karin M, Lin A (2002) NF-κB at the crossroads of life and death. Nat Immunol 3(3):221–227
129. Nguyen DP et al (2014) Recent insights into NF-kappaB signalling pathways and the link between inflammation and prostate cancer. BJU Int 114(2):168–176
130. Suh J et al (2002) Mechanisms of constitutive NF-κB activation in human prostate cancer cells. Prostate 52(3):183–200
131. Shukla S et al (2004) Nuclear factor-κB/p65 (Rel a) is constitutively activated in human prostate adenocarcinoma and correlates with disease progression. Neoplasia 6(4):390–400
132. Fuxe, J. and M.C.I. Karlsson, Epithelial–Mesenchymal transition: a link between cancer and inflammation, in cancer and inflammation mechanisms : chemical, biological, and clinical aspects, Y. U. Hiraku, S. O. Kawanishi, and H. Oshima, Editors. p. 1 online resource (409 pages)
133. Heldin CH, Landstrom M, Moustakas A (2009) Mechanism of TGF-beta signaling to growth arrest, apoptosis, and epithelial-mesenchymal transition. Curr Opin Cell Biol 21(2):166–176
134. Xu J, Lamouille S, Derynck R (2009) TGF-beta-induced epithelial to mesenchymal transition. Cell Res 19(2):156–172
135. Wu D, Pan W (2010) GSK3: a multifaceted kinase in Wnt signaling. Trends Biochem Sci 35(3):161–168
136. Buijs JT et al (2007) TGF-beta and BMP7 interactions in tumour progression and bone metastasis. Clin Exp Metastasis 24(8):609–617
137. Fuxe J, Vincent T, Garcia A (2010) De Herreros, Transcriptional crosstalk between TGF-beta and stem cell pathways in tumor cell invasion: role of EMT promoting Smad complexes. Cell Cycle 9(12):2363–2374
138. Maitah MY et al (2011) Up-regulation of sonic hedgehog contributes to TGF-beta1-induced epithelial to mesenchymal transition in NSCLC cells. PLoS One 6(1):e16068
139. Oft M et al (1996) TGF-beta1 and ha-Ras collaborate in modulating the phenotypic plasticity and invasiveness of epithelial tumor cells. Genes Dev 10(19):2462–2477

140. Zavadil J et al (2004) Integration of TGF-beta/Smad and Jagged1/notch signalling in epithelial-to-mesenchymal transition. EMBO J 23(5):1155–1165
141. Massague J (2008) TGFbeta in Cancer. Cell 134(2):215–230
142. Dalal BI, Keown PA, Greenberg AH (1993) Immunocytochemical localization of secreted transforming growth factor-beta 1 to the advancing edges of primary tumors and to lymph node metastases of human mammary carcinoma. Am J Pathol 143(2):381–389
143. Kawada M et al (2008) Transforming growth factor-beta1 modulates tumor-stromal cell interactions of prostate cancer through insulin-like growth factor-I. Anticancer Res 28(2A):721–730
144. Bates RC, Mercurio AM (2003) Tumor necrosis factor-alpha stimulates the epithelial-to-mesenchymal transition of human colonic organoids. Mol Biol Cell 14(5):1790–1800
145. Wu Y et al (2009) Stabilization of snail by NF-kappaB is required for inflammation-induced cell migration and invasion. Cancer Cell 15(5):416–428
146. Wyckoff J et al (2004) A paracrine loop between tumor cells and macrophages is required for tumor cell migration in mammary tumors. Cancer Res 64(19):7022–7029
147. Jordan CT, Guzman ML, Noble M (2006) Cancer stem cells. N Engl J Med 355(12):1253–1261
148. Clarke MF et al (2006) Cancer stem cells--perspectives on current status and future directions: AACR workshop on cancer stem cells. Cancer Res 66(19):9339–9344
149. Tanno T, Matsui W (2014) In: Hiraku YU, Kawanishi SO, Oshima H (eds) Stem cell theory and inflammation-related cancer, in cancer and inflammation mechanisms : Chemical, biological, and clinical aspects, p 1 online resource (409 pages)
150. Widera D et al (2006) Tumor necrosis factor alpha triggers proliferation of adult neural stem cells via IKK/NF-kappaB signaling. BMC Neurosci 7:64
151. Audet J et al (2001) Distinct role of gp130 activation in promoting self-renewal divisions by mitogenically stimulated murine hematopoietic stem cells. Proc Natl Acad Sci U S A 98(4):1757–1762
152. Tothova Z et al (2007) FoxOs are critical mediators of hematopoietic stem cell resistance to physiologic oxidative stress. Cell 128(2):325–339
153. Ito K et al (2007) Regulation of reactive oxygen species by Atm is essential for proper response to DNA double-strand breaks in lymphocytes. J Immunol 178(1):103–110
154. Ito K et al (2006) Reactive oxygen species act through p38 MAPK to limit the lifespan of hematopoietic stem cells. Nat Med 12(4):446–451
155. Liu Y et al (2009) p53 regulates hematopoietic stem cell quiescence. Cell Stem Cell 4(1):37–48
156. Adelman DM, Maltepe E, Simon MC (1999) Multilineage embryonic hematopoiesis requires hypoxic ARNT activity. Genes Dev 13(19):2478–2483
157. Gilbertson RJ, Rich JN (2007) Making a tumour's bed: glioblastoma stem cells and the vascular niche. Nat Rev Cancer 7(10):733–736
158. Guzman ML et al (2005) The sesquiterpene lactone parthenolide induces apoptosis of human acute myelogenous leukemia stem and progenitor cells. Blood 105(11):4163–4169
159. Sullivan R, Graham CH (2008) Chemosensitization of cancer by nitric oxide. Curr Pharm Des 14(11):1113–1123
160. Tiligada E (2006) Chemotherapy: induction of stress responses. Endocr Relat Cancer 13(Suppl 1):S115–S124
161. Pervaiz S, Clement MV (2004) Tumor intracellular redox status and drug resistance--serendipity or a causal relationship? Curr Pharm Des 10(16):1969–1977
162. Zavadil J, Bottinger EP (2005) TGF-beta and epithelial-to-mesenchymal transitions. Oncogene 24(37):5764–5774
163. Vesuna F et al (2009) Twist modulates breast cancer stem cells by transcriptional regulation of CD24 expression. Neoplasia 11(12):1318–1328
164. Gupta PB et al (2009) Identification of selective inhibitors of cancer stem cells by high-throughput screening. Cell 138(4):645–659

165. Mani SA et al (2008) The epithelial-mesenchymal transition generates cells with properties of stem cells. Cell 133(4):704–715
166. Nguyen QD et al (2005) Commutators of PAR-1 signaling in cancer cell invasion reveal an essential role of the rho-rho kinase axis and tumor microenvironment. Oncogene 24(56):8240–8251
167. Yang MH et al (2008) Direct regulation of TWIST by HIF-1alpha promotes metastasis. Nat Cell Biol 10(3):295–305
168. Merchant AA, Matsui W (2010) Targeting hedgehog--a cancer stem cell pathway. Clin Cancer Res 16(12):3130–3140
169. Jamieson CH et al (2004) Granulocyte-macrophage progenitors as candidate leukemic stem cells in blast-crisis CML. N Engl J Med 351(7):657–667

Chapter 4
The Impact of Centrosome Pathologies on Prostate Cancer Development and Progression

Heide Schatten and Maureen O. Ripple

Abstract The significant role of centrosomes in cancer cell proliferation has been well recognized (reviewed in Schatten H, Histochem Cell Biol 129:667–86 (2008); Schatten H, Sun Q-Y, Microsc Microanal 17(4):506–512 (2011); Schatten H, Sun Q-Y, Reprod Fertil Dev. https://doi.org/10.1071/RD14493 (2015a); Schatten H, Sun Q-Y, Centrosome-microtubule interactions in health, disease, and disorders. In: Schatten H (ed) The cytoskeleton in health and disease. Springer Science+Business Media, New York (2015b)) and new research has generated new interest and new insights into centrosomes as potential targets for cancer-specific therapies. The centrosome is a key organelle serving multiple functions through its primary functions as microtubule organizing center (MTOC) that is also an important communication center for processes involved in cellular regulation; transport to and away from centrosome-organized microtubules along microtubules is essential for cellular activities including signal transduction and metabolic activities. New research on cancer cell centrosomes has generated new insights into centrosome dysfunctions in cancer cells in which centrosome phosphorylation, balance of centrosomal proteins, centrosome regulation and duplication are impaired. Among the hallmarks of cancer cells are multipolar spindles or abnormal bipolar spindles that are formed as a result of centrosome protein expression imbalances, abnormalities in centrosome structure and abnormalities in clustering of centrosomal components that are critical for bipolar mitotic apparatus formation. Centrosome abnormalities in cancer cells can be the result of multiple factors including environmental influences and toxicants that can affect centrosome functions by inducing centrosome pathologies leading to abnormal cancer cell proliferation. These topics are addressed in this review with focus on prostate-specific therapy strategies to target centrosome abnor-

H. Schatten (✉)
Department of Veterinary Pathobiology, University of Missouri, Columbia, MO, USA
e-mail: SchattenH@missouri.edu

M. O. Ripple
Dartmouth-Hitchcock/Geisel School of Medicine Office of Development
& Geisel School Alumni Relations, Hanover, NH, USA
e-mail: maureen.ripple@hitchcock.org; maureen.o.ripple@dartmouth.edu

© Springer International Publishing AG, part of Springer Nature 2018   67
H. Schatten (ed.), *Cell & Molecular Biology of Prostate Cancer*,
Advances in Experimental Medicine and Biology 1095,
https://doi.org/10.1007/978-3-319-95693-0_4

malities. We will also address loss of cell polarity in cancer cells in which centrosome dysfunctions play a role as well as the loss of primary cilia in prostate cancer development and progression.

Keywords Centrosomes · Cancer cell proliferation · Cancer-specific therapies · Microtubule organizing center (MTOC) · Microtubules · Centrosome dysfunctions · Multipolar spindles · Centrosome clustering · Centrosome pathologies · Centrosome abnormalities · Cell polarity · Primary cilia · Prostate cancer development and progression

4.1   Introduction

Advances in cell and molecular biology as well as genetic manipulations have allowed new approaches to gain new insights into centrosome biology and control abnormal centrosome behavior and function. While best known for their microtubule nucleating and organizing capabilities (MTOCs) centrosomes are highly dynamic structures that carry out multiple functions including formation of the bipolar spindle during mitosis and cell division to separate chromosomes accurately to the dividing daughter cells; organization of microtubules for maintenance of cell shape; reorganization of microtubules during cellular polarization; transport of mitochondria to their functional destinations; docking station for enzyme-carrying vesicles that are translocated along microtubules; and a diversity of others. In addition, centrosomes serve as central hub for signal transduction molecules, thereby playing a role in signaling pathways that are critical for cell cycle regulation.

Abnormal phosphorylation of centrosomes can result in nucleation of abnormally increased numbers of microtubules, formation of multipolar mitosis or bipolar mitosis with abnormal amounts of centrosomal proteins with consequences for abnormal separation of chromosomes and aneuploidy associated with genomic instability. We do not yet understand when centrosome changes occur that play a role in cancer development and progression and we do not yet understand the relationships between cause and effect but we have some indications that the interrelationships are cumulative and can result in vicious cycles of no return to normalcy. We know that cancer cell centrosomes are significantly different from non-cancer cell centrosomes including in their state of phosphorylation [45]; in addition to being hyper-phosphorylated in mitosis cancer-cell centrosomes are also phosphorylated in interphase while they are mainly phosphorylated in mitosis in regular cell cycles. The loss of phosphorylation control in cancer cells involves several key kinases that are involved in the transition from G2 to mitosis [3, 18, 19, 61] and play a role in centrosome protein phosphorylation while dephosphorylation takes place when cells exit mitosis which is important for regulated centrosome functions. Other factors involved in cancer cell centrosome dysregulation include disruption of centrosome duplication [34], DNA damage caused by radiation [64], protein degradation dysfunctions [18, 30,

Fig. 4.1 A typical mammalian centrosome is composed of two centrioles surrounded by a meshwork of proteins embedded in a matrix called the pericentriolar material (PCM). Gamma-tubulin and the gamma- tubulin ring complex that nucleate microtubules along with associated proteins are embedded in the PCM. Highlighted in this diagram are two centrosomal complexes, the microtubule nucleating complex and the microtubule anchoring complex. Adapted from Schatten [71]

61], and effects of environmental factors [56, 74, 94]. For prostate cancer, the effects of bisphenol A have been studied in human prostate cancer and have been correlated with early-onset prostate cancer and centrosome amplification [89]. The specific effects of endocrine disruptors on prostate cancer will be addressed below in Sect. 4.3. The present review will address (1) structure, function, and regulation of centrosomes and abnormalities in prostate cancer, (2) the role of primary cilia and signaling through primary cilia in prostate cancer, (3) toxicants that affect centrosomes with consequences for prostate cancer development and progression, and (4) centrosomes as target for prostate cancer therapy and prevention.

(1) Structure, function, and regulation of centrosomes in epithelial cells and abnormalities in prostate cancer cells

The structure of somatic cell centrosomes has been described in previous reviews [71, 75–77] and is only briefly addressed in the present review. A typical mammalian somatic cell centrosome is composed of a centrally positioned perpendicularly oriented centriole pair that is surrounded by a centrosomal matrix (Fig. 4.1), often-times also referred to as pericentriolar material (PCM) composed of a lattice of coiled-coil proteins. This centrosomal matrix contains numerous specific centrosomal proteins including the γ-tubulin ring complexes (γ-TuRCs), pericentrin, centrin, and calcium-sensitive fibers (Salisbury [66]; reviewed in Schatten [71]). The centrosome organelle is not membrane bound which allows direct interactions with cytoplasmic components facilitated by microtubules that are dynamically organized and reorganized by centrosomes throughout the cell cycle.

Remodeling of centrosomes throughout the cell cycle includes remodeling by specific centrosome proteins to carry out cell cycle-specific functions while centrioles do not undergo similar reorganizations except that they duplicate during the S-phase in a semiconservative duplication process during which a younger (daughter) centriole forms perpendicular to the older (mother) centriole (Fig. 4.1). Daughter and mother centrioles in mammalian somatic cells are composed of nine outer triplet microtubules forming a small tube that does not contain central microtubules. The distinction between daughter and mother centrioles becomes important considering that there are structural and functional differences in that mother centrioles contain appendages and serve as seed structures for the formation of non-motile primary cilia that are formed as single cilia at the surface of epithelial cells as well as most other cells in the human body. The specifics of primary cilia in prostate cancer cells will be discussed in Sect. 4.2. Centrioles within the centrosome complex are important for the assembly of specific centrosome proteins and for the duplication of centrosomal material [67]. The nature of the centrosome matrix is still not well understood although it is known that numerous centrosomal proteins are associated with the centrosome matrix that undergo cell cycle-specific regulation. The amount and composition of centrosome proteins within the centrosome matrix varies and includes centrosome core proteins that are permanently associated with the centrosome structure while others are part of the cell cycle-dependent structural centrosomal changes in most cell systems. The amount and composition of centrosome proteins within the centrosome matrix is precisely regulated during normal cell cycles but becomes deregulated during cancer development and progression in that some of the centrosomal proteins are overexpressed and play a role in centrosome amplification [5, 20]. A great number of centrosomal proteins have been identified in purified centrosomes by mass spectrometric analysis and include structural proteins (alpha-tubulin, beta-tubulin, γ-tubulin, γ-tubulin complex components 1–6, centrin 2 and 3, AKAP450, pericentrin/kendrin, ninein, pericentriolar material 1 (PCM1), ch-TOG protein, C-Nap1, Cep250, Cep2, centriole-associated protein CEP110, Cep1, centriolin, centrosomal P4.1-associated protein (CPAP), CLIP-associating proteins CLASP1 and CLASP 2, ODF2, cenexin, Lis1, Nudel, EB1, centractin, myomegalin); regulatory molecules (cell division protein 2 (Cdc2), Cdk1, cAMP-dependent protein kinase type II-alpha regulatory chain, cAMP-dependent protein kinase-alpha catalytic subunit, serine/threonine protein kinase Plk1, serine/threonine protein kinase Nek2, serine/threonine protein kinase Sak, Casein kinase I, delta and epsilon isoforms, protein phosphatase 2A, protein phosphatase 1 alpha isoform, 14–3-3 proteins, epsilon and gamma isoforms); motor and motor-related proteins (dynein heavy chain, dynein intermediate chain, dynein light chain, dynactin 1, p150 Glued, dynactin 2, p50, dynactin 3); and the heat shock proteins, heat shock protein Hsp90, TCP subunits, and heat shock protein Hsp73.

The γ-tubulin ring complex, pericentrin, centrin, and the centrosome-associated protein NuMA (Nuclear Mitotic Apparatus protein) have been discussed in more detail in previous reviews [75–77]. Briefly, γ-tubulin is primarily found in the centrosome matrix core structure, but it can also be a microtubule nucleating protein in areas away from the centrosome which is important for polarized epithelial cells in which γ-tubulin nucleates interphase microtubules that are needed for cellular communication between the apical and baso-lateral membranes of polarized epithelial cells. The microtubule minus-end anchoring protein, ninein [55], has a major role in microtubule anchorage at centrosomes as well as at non-centrosomal anchorage sites. The centrosomal protein pericentrin depends on dynein for assembly onto centrosomes [95]. It plays a role in centrosome and spindle organization [12, 15, 95] and forms a ca. 3-MDa complex with γ-tubulin. Pericentrin is involved in recruiting γ-tubulin to centrosomes [12], and it is part of the pericentrin/AKAP450 centrosomal targeting (PACT) domain [21]. Centrins are primarily associated with centrioles and are an intrinsic component of centrosomes with an essential role in the duplication of centrosomes ([39, 47, 65, 67]; reviewed in Manandhar et al. [48]). Loss of cell polarity is a hallmark of cancer cells but the process involving loss of cellular polarization is still not well understood. Loss of microtubule organization by γ-tubulin and anchoring proteins at the apical cell surface may play a role in this process.

The centrosome-associated protein NuMA deserves closer attention, as it forms an insoluble crescent around the centrosome area facing toward the central mitotic spindle with the important function to cross-link spindle microtubules and tether microtubules precisely into the bipolar mitotic apparatus (Fig. 4.1) [50]. This multifunctional protein (reviewed in Sun and Schatten [86, 87]) serves as nuclear matrix protein in the nucleus during interphase but does not associate with the interphase centrosome. It becomes an important centrosome-associated protein when it moves out of the nucleus during nuclear envelope breakdown and disperses into the cytoplasm to associate with microtubules for translocation to the centrosomal area in a dynein/dynactin-mediated process. Cdk1/cyclin B-dependent phosphorylation is important for translocation of NuMA from the nucleus to the cytoplasm (Saredi et al. [68]; reviewed in Sun and Schatten [86, 87]) and again for NuMA's dissociation from the centrosomal area during exit from mitosis. NuMA dysfunctions are associated with mitotic dysfunctions in several cell systems (reviewed in [1, 86]) and play a role in mitotic abnormalities in cancer cells [31], as reported for breast cancer.

(2) The role of primary cilia in cellular regulation and abnormalities in prostate cancer

The past decade has brought us significant new insights into the important role of primary cilia for signal transduction and cell cycle regulation and its impact on cancer development and progression. One single primary cilium protrudes from almost all cells in our body and plays a significant role in cellular and cell cycle communication (Fig. 4.2; reviewed in Schatten [72]). Furthermore, we now know the intimate and tight correlation of the primary cilia-centrosome cycle that has been clearly established (reviewed in Pan and Snell [58]; reviewed in Schatten and Sun [75]). The mother centriole within the centrosome complex serves as the seed for primary cilia formation when during G1 the distal end of the mother centriole becomes associated with a membrane vesicle (reviewed by Pan and Snell [58]) that grows into a ciliary vesicle, surrounds a forming axoneme, and fuses with the plasma membrane during primary cilia formation. Centrioles duplicate during the subsequent S phase and the primary cilium lengthens and achieves the mature length during the G2

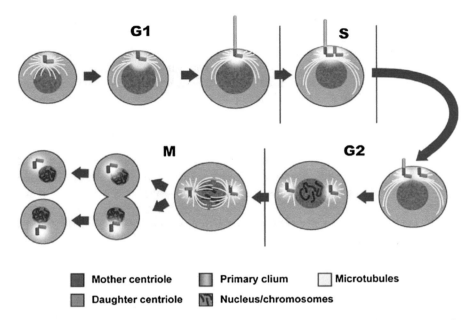

Fig. 4.2 Schematic representation of primary cilium—cell cycle relationships The mother centriole undergoes docking to membrane in G1 while accessory proteins build the ciliary axoneme from the mother centriole's triplet microtubules. Cell centriole and DNA replication start during S, and centriole maturation occurs in G2. During mitosis (M) centrioles participate in mitotic spindle formation. The primary cilium reassembles in G1. Adapted from Schatten and Sun [75]

phase. Centriole and primary cilium shortening then takes place at the G2/M transition and mitotic cells do not display primary cilia (reviewed in [40, 72, 73]).

Unlike motile cilia the primary cilium is a non-motile single cilium composed of 9 outer microtubule doublets with no central microtubule pair ("9 + 0") [9, 90, 92]. It is covered by a specialized receptor-rich plasma membrane that is critically important for communicating signals from the external cellular environment to its associated cell body (reviewed by Li and Hu [40]). Primary cilia dysfunctions are associated with numerous diseases including polycystic kidney syndrome and other diseases or disorders for which the cell, molecular, and genetic aspects of primary cilia dysfunctions have been reported. Details of signal transduction cascades between primary cilia and the centrosome and their essential role for accurate cell cycle progression have been reported and discussed ([9, 10, 24, 53, 62, 69, 71, 90]; reviewed in Li and Hu [40]).

At least three major pathways require signaling through primary cilia and include the Wnt, hedgehog, and platelet-derived growth factor (PDGF) pathways [2, 84]. In addition, MAP kinase signaling between primary cilia and centrosomes is important for centrosome functions which includes downstream signaling cascades such as phosphorylation and activation of the Akt and Mek1/2–Erk1/2 pathways [82]. To communicate signals received by the primary cilium to the cell body an intraflagellar transport (IFT) system is essential (reviewed in Li and Hu [40]).

In cancer cells, severe primary cilia dysfunctions have been reported which in part are related to signal dysfunctions, and to cellular, molecular and genetic disorders that play a role in cancer development and progression. In advanced tumors, the primary cilium becomes dislodged from the cell surface when cellular polarization is lost which results in the basal body of the primary cilium being translocated into the cell body [43, 44, 80]. This in turn can result in forming seeds for microtubule nucleation and organization that may participate in additional spindle formations, thereby participating in the abnormal mitotic process that may lead to increases in aneuploidy and abnormal cell divisions.

Genetic factors play a role in primary cilia dysfunctions and several primary cilia-associated genes are known to be mutated in cancers which includes Gli1, DNAH9, and RPGR1P1 (reviewed in Schatten and Sun [75]). An important oncogenic kinase is Aurora A kinase (Aurora A) that is localized to the basal body of primary cilia. It may play a role in primary cilia disassembly or may block primary cilia reassembly in coordination with other interacting proteins [27]. Aurora A may be dysfunctional in cancer cells associated with primary cilia–cell cycle dysfunctions.

As mentioned above, primary cilia have an essential role in hedgehog signaling which is among the best studied signaling pathways associated with primary cilia-cellular communications. The transmembrane protein, Smoothened (Smo), in the primary cilium plays a role in the activation of the hedgehog pathway; subsequent hedgehog-dependent transcription is mediated by the three transcription factors, Gli1, Gli2, and Gli3 [23, 32]. The role for hedgehog signaling through primary cilia in cancer progression has been reported and interfering with the hedgehog pathways has been proposed as strategy for the design of new therapeutics [22, 85].

In human prostate cancer Hassounah et al. [22] reported a decrease in the percentage of ciliated cells in prostatic intraepithelial neoplasia (PIN), invasive cancer and perineural invasion lesions when compared to normal prostate tissue. They also observed shorter cilia in PIN, cancer, and perineural invasion lesions which may affect function. When analyzing the Wnt signaling pathway the authors found that primary cilia normally function to suppress the Wnt signaling pathway in epithelial cells and that cilia loss may play a role in increased Wnt signaling in some prostate cancers which may open up new targeted treatment strategies related to Wnt signaling.

Other strategies for therapeutic intervention of abnormal signaling cascades in cancer have been proposed and include the PDGF signaling pathway by primary cilia that has been well explored by Schneider et al. [82]. In breast cancer, it has been shown that expression of PDGFRα is a poor prognostic indicator of breast cancer [6, 28], clearly linking PDGF signaling through primary cilia to breast cancer. In prostate cancer, PDGFRα has been shown to play a role in survival and growth of prostate cancer cells in the bone, supporting early metastatic foci [46]. More data are needed to investigate PDGF signaling pathway by primary cilia in prostate cancer.

(3) Toxicants that affect centrosomes with consequences for prostate cancer development and progression

The susceptibility of centrosomes to drugs and toxic agents has been reported for meiotic spindles in mammalian oocytes (reviewed in Miao et al. [51]; and references therein) which can cause aneuploidy that has been implicated in infertility, developmental abnormalities and early childhood cancer [51, 52]. In these studies much attention has been paid to bisphenol-A (BPA), an alkylphenol and environmental estrogen-like chemical with weak estrogenic activity that affects centrosome and spindle integrity in MI and MII spindles of mice [4]. These studies showed that BPA causes a time- and dose-dependent delay in cell cycle progression, primarily by interfering with centrosomal proteins that may be degraded by BPA. Other studies added information confirming the toxic effects of BPA on centrosomes in reproductive cells ([17, 57]; reviewed in Miao et al. [51]). While BPA is not classified as a carcinogen a direct relationship between BPA and increased prostate cancer has recently been observed [89]. In this study on 60 urology patients it was suggested that low levels of BPA exposure correlates with early-onset prostate cancer and promotes centrosome amplification and anchorage-independent growth in vitro. BPA had been used extensively in thousands of consumer products to produce polycarbonate plastic and epoxy resins. Exposure to BPA has been reported to be absorbed by skin, and it causes effects through inhalation, and ingestion from contaminated food and water. Use of BPA has been reduced in recent years based on animal studies that showed damaging effects to reproductive systems, the immune system as well as being a disruptor of metabolism and because of implications in cardiovascular disease, obesity, diabetes, and others [14, 16, 37]. In vitro and animal studies have further shown that BPA exposure is significantly implicated in prostate cancer [26, 29, 33, 60, 88]. The studies by Tarapore et al. [89] were the first to show a correlation of BPA to prostate cancer in humans and that the effect of BPA is on the centrosome cycle contributing to prostate carcinogenesis.

Despite not being recognized as a carcinogen the effects of BPA on centrosomes leading to prostate cancer points to a contributing factor although further studies are needed to determine with clarity the cause and effect relationships. The relationship between cause and effect concerning centrosome pathologies in cancer has been debated in numerous papers (reviewed in Schatten [71–73]) and it is certain that centrosome amplification is a hallmark of cancer although several factors may play a role in centrosome abnormalities and cancer development and progression. The study by Tarapore et al. [89] showed clearly that treatment with BPA increased the number of cells with abnormal centrosomes and that BPA plays a role in prostate cancer development and disease progression.

(4) Centrosomes as targets for prostate cancer therapy

Because centrosome pathologies are associated with prostate cancer and other cancers it has been proposed to target centrosomes for the development of specific therapies that affect cancer cell centrosomes while not affecting centrosomes in non-cancer cells. Several avenues have been proposed and include direct or indirect targeting of centrosomes or targeting via interfering with centrosome-related signal transduction pathways that are abnormal in cancer cells. As mentioned above, centrosomes are abnormally phosphorylated in cancer cells. Several centrosome-phosphorylating kinases are overexpressed in cancer cells which includes polo-like kinases, cyclin-dependent kinases, Aurora kinases, and several others (reviewed in [3, 19, 71]) that have advanced to some early clinical trials [8, 83]. In addition, histone deacetylases like HDAC1, HDAC5, and SIRT1 have also been explored for their targeting potential, as they inhibit centrosome duplication and amplification [42]. Aryl hydrocarbon receptor agonists are also being considered as therapeutic targets to inhibit centrosome pathologies, as they affect centriole overduplication which is important considering that centrioles are involved in centrosome duplication. Targeting aryl hydrocarbon receptor signaling in the centrosome cycle [7, 35] will therefore target centrosome amplification in cancer cells.

As outlined in Chap. 1 of this book taxol is a primary drug to control castration-resistant prostate cancer. Paclitaxel (or other taxol derivatives) is known to target microtubule dynamics by primarily inhibiting depolymerization of microtubules, thereby preventing progression of mitosis and cell division [78, 81]. In addition, taxol has been reported to interact with microtubules at the centrosome–microtubule nucleation sites [11, 13], thereby affecting the capacity of centrosomes to nucleate microtubules in taxol-treated cells [11].

Several newer promising approaches to target centrosomes to inhibit cancer cell proliferation involves inhibition of centrosome clustering that has briefly been mentioned above. The process of centrosome clustering has previously been reviewed in detail [36, 93] and proteins required for centrosome clustering have been determined [38].

As mentioned above, multipolar cells containing multiple centrosomes are hallmarks of cancer, but cancer cells have developed a mechanism to suppress multipolarity by clustering their extra centrosomes into pseudo-bipolar spindles, thereby allowing cancer cell survival with intrinsic centrosomal abnormalities that may be manifested in subsequent cell cycles.

Several antimitotic drugs have been explored for their effects on microtubules and centrosomes in cancer cells. The antimitotic drug griseofulvin arrests cells at the G2/M transition stage in a concentration-dependent manner and it has recently been shown that griseofulvin specifically inhibits clustering of supernumerary centrosomes in cancer cells (reviewed in Krämer et al. [36]). As supernumerary centrosomes are hallmarks of cancer cells (reviewed in Schatten [71]) this drug may be utilized to prevent the formation of amplified centrosome clustering into an abnormal bipolar mitotic apparatus, thereby causing cancer cells to undergo fragmentation rather than abnormal cell division because centrosome clustering is prevented. Previous studies already had shown that griseofulvin induces multipolar mitoses in tumor cells [25, 59, 63, 70] and is therefore a good candidate for further development into an optimal applicable drug and subsequent clinical trials. The rationale behind developing anti-centrosome-clustering therapies comes from our knowledge that non-clustered centrosomes cannot form a bipolar mitotic apparatus with amplified centrosomal components but induce cell death following cell fragmentation rather than allowing formation of aneuploid cells with consequences for genomic instability. This approach is attractive, as centrosome clustering pathways are dispensable in cells with normal centrosome numbers, but centrosome clustering is required for supernumerary centrosomes to form a bipolar mitotic apparatus with amplified centrosomes that can undergo cell division but result in subsequent cellular and tissue abnormalities. Griseofulvin is already approved as an effective orally administered antifungal drug that affects microtubule functions in vivo and in vitro ([49, 70, 79, 91]; reviewed in Schatten [71]). At present it is not clear how griseofulvin specifically affects centrosome clustering; it may interfere with microtubule minus ends rather than interacting with the centrosome structure directly. Investigating the mechanism by which griseofulvin inhibits centrosome clustering will be important to determine combination therapies to target certain subpopulations of cells in cancer tissue.

One specific approach to prevent centrosome clustering in cancer cells has emerged in recent years and that is targeting KIFC1, a kinesin-like protein (kinesin motor) that plays a critical role in clustering the multiple centrosomes that are specific for cancer cells. It has been shown that KIFC1 is non-essential in normal somatic cells and presents a highly suitable target to control clustering of centrosomes into abnormal bipolar spindles in cancer cells. One consideration needs to be taken into account and that is that KIFC1 also plays a role in certain vesicular and organelle trafficking, spermiogenesis, oocyte development, embryo gestation, and double-strand DNA transportation (reviewed in Xiao and Yang [93]). These concerns regarding possible side effects may outweigh the potentially significant treatment possibilities by targeting KIFC1.

KIFC1 has been reported to be upregulated in breast cancer and particularly in estrogen receptor negative, progesterone receptor negative and triple negative breast cancer [41] while it is absent in normal human mammary epithelial cells. Inhibition of KIFC1 resulted in anti-breast cancer activity. KIFC1 has also been implicated in serous ovarian adenocarcinomas [54] and it has been shown to predict aggressive disease course, therefore serving as a biomarker to predict the aggressiveness of the disease [54].

4.2   Conclusion and Future Directions

Significant progress has been made in the overall diagnosis and treatment of prostate cancer including personalized treatment options to target specific abnormalities in different stages of prostate cancer. New therapeutic approaches are needed to target advanced stages of prostate cancer. While androgen deprivation is effective in early states of disease development taxol is being used most frequently when androgen deprivation becomes ineffective. Taxol is being used as an effective drug but drug resistance may occur and different treatment strategies are needed. Targeting centrosomes to inhibit abnormal cell proliferation is one approach to pursue new treatment strategies and may involve the antimitotic drug griseofulvin to inhibit centrosome clustering and prevent abnormal cell proliferation by eliciting cell fragmentation followed by cell death. Targeting KIFC1 is another approach to prevent centrosome clustering in cancer cells, as KIFC1 is primarily present in cancer but not in non-cancerous epithelial cells. KIFC1 is not needed and dispensable in non-cancerous epithelial cells. These treatments may prove effective in eliminating cancer cells while not affecting non-cancer cells that do not rely on centrosome clustering. Both treatments will prevent cancer cells from abnormal cell divisions and affect their viability. More research is still needed on griseofulvin and KIFC1 for practical applications to treat prostate cancer to specifically target cancer cells without causing side effects in non-cancer cells.

References

1. Alvarez Sedó CA, Schatten H, Combelles C, Rawe VY (2011) The nuclear mitotic apparatus protein NuMA: localization and dynamics in human oocytes, fertilization and early embryos. Mol Hum Reprod 17(6):392–398. https://doi.org/10.1093/molehr/gar009
2. Berbari NF, O'Connor AK, Haycraft CJ, Yoder BK (2009) The primary cilium as a complex signaling center. Curr Biol 19:R526–R535
3. Boutros R (2012) Regulation of centrosomes by cyclin-dependent kinases. In: Schatten H (ed) The centrosome, Chap 11. Springer Science and Business Media, New York
4. Can A, Semiz O, Cinar O (2005) Bisphenol-a induces cell cycle delay and alters centrosome and spindle microtubular organization in oocytes during meiosis. Mol Hum Reprod 11:389–396
5. Carroll E, Okuda M, Horn HF, Biddinger P, Stambrook PJ, Gleich LL, Li YQ, Tarapore P, Fukasawa K (1999) Centrosome hyperamplification in human cancer: chromosome instability induced by p53 mutation and/or Mdm2 overexpression. Oncogene 18:1935–1944
6. Carvalho I, Milanezi F, Martins A, Reis RM, Schmitt F (2005) Overexpression of platelet-derived growth factor receptor alpha in breast cancer is associated with tumour progression. Breast Cancer Res 7:R788–R795
7. Chan JY (2011) A clinical overview of centrosome amplification in human cancers. Int J Biol Sci 7:1122–1144
8. Cheung CH, Coumar MS, Chang JY, Hsieh HP (2011) Aurora kinase inhibitor patents and agents in clinical testing: an update (2009–10). Expert Opin Ther Pat 21:857–884
9. D'Angelo A, Franco B (2009) The dynamic cilium in human diseases. PathoGenetics 2(3):1–15
10. Davenport JR, Yoder BK (2005) An incredible decade for the primary cilium: a look at a once-forgotten organelle. Am J Physiol Renal Physiol 289:F1159–F1169
11. De Brabander M, Geuens G, Nuydens R, Willebrords R, De Mey J (1981) Taxol induces the assembly of free microtubules in living cells and blocks the organizing capacity of the centrosomes and kinetochores. Proc Natl Acad Sci U S A 78:5608–5612
12. Dictenberg J, Zimmerman W, Sparks C, Young A, Vidair C, Zheng Y, Carrington W, Fay F, Doxsey SJ (1998) Pericentrin and gamma tubulin form a protein complex and are organized into a novel lattice at the centrosome. J Cell Biol 141:163–174
13. Dimitriadis I, Katsaros C, Galatis B (2001) The effect of taxol on centrosome function and microtubule organization in apical cells of Sphacelaria rigidula (Phaeophyceae). Phycol Res 49:23–34
14. Donohue KM, Miller RL, Perzanowski MS, Just AC, Hoepner LA et al (2013) Prenatal and postnatal bisphenol a exposure and asthma development among inner-city children. J Allergy Clin Immunol 131:736–742

15. Doxsey SJ, Stein P, Evans L, Calarco P, Kirschner M (1994) Pericentrin, a highly conserved protein of centrosomes involved in microtubule organization. Cell 76:639–650
16. Ehrlich S, Williams PL, Missmer SA, Flaws JA, Ye X et al (2012) Urinary bisphenol a concentrations and early reproductive health outcomes among women undergoing IVF. Hum Reprod 27:3583–3592
17. Eichenlaub-Ritter U, Vogt E, Cukurcam S, Sun F, Pacchierotti F, Parry J (2008) Exposure of mouse oocytes to bisphenol a causes meiotic arrest but not aneuploidy. Mutat Res 651:82–92
18. Fisk HA (2012) Many pathways to destruction: the centrosome and its control by and role in regulated proteolysis. In: Schatten H (ed) The centrosome, Chap 8. Springer Science and Business Media, New York
19. Fukasawa K (2012) Molecular links between centrosome duplication and other cell cycle associated events. In: Schatten H (ed) The centrosome, Chap 10. Springer Science and Business Media, New York
20. Fukasawa K, Choi T, Kuriyama R, Rulong S, Vande Woude GF (1996) Abnormal centrosome amplification in the absence of p53. Science 271:1744–1747
21. Gillingham AK, Munro S (2000) The PACT domain, a conserved centrosomal targeting motif in the coiled-coil proteins AKAP450 and pericentrin. EMBO Rep 1:524–529
22. Hassounah NB, Bunch TA, McDermott KM (2012) Molecular pathways: the role of primary cilia in Cancer progression and therapeutics with a focus on hedgehog signaling. Clin Cancer Res 18(9):2429–2435
23. Haycraft CJ, Banizs B, Aydin-Son Y et al (2005) Gli2 and gli3 localize to cilia and require the intraflagellar transport protein polaris for processing and function. PLoS Genet 1:e53
24. Hildebrandt F, Otto E (2005) Cilia and centrosomes: a unifying pathogenic concept for cystic kidney disease? Nat Rev Genet 6:928–940
25. Ho YS, Duh JS, Jeng JH, Wang YJ, Liang YC, Lin CH, Tseng CJ, Yu CF, Chen RJ, Lin JK (2001) Griseofulvin potentiates antitumorigenesis effects of nocodazole through induction of apoptosis and G2/M cell cycle arrest in human colorectal cancer cells. Int J Cancer 91:393–401
26. Ho SM, Tang WY, de Belmonte FJ, Prins GS (2006) Developmental exposure to estradiol and bisphenol a increases susceptibility to prostate carcinogenesis and epigenetically regulates phosphodiesterase type 4 variant 4. Cancer Res 66:5624–5632
27. Inoko A, Matsuyama M, Goto H, Ohmuro-Matsuyama Y, Hayashi Y, Enomoto M, Ibi M, Urano T, Yonemura S, Kiyono T, Izawa I, Inagaki M (2012) Trichoplein and aurora a block aberrant primary cilia assembly in proliferating cells. J Cell Biol 197(3):391–405
28. Jechlinger M, Sommer A, Moriggl R et al (2006) Autocrine PDGFR signaling promotes mammary cancer metastasis. J Clin Invest 116:1561–1570
29. Jenkins S, Wang J, Eltoum I, Desmond R, Lamartiniere CA (2011) Chronic oral exposure to bisphenol a results in a nonmonotonic dose response in mammary carcinogenesis and metastasis in MMTV-erbB2 mice. Environ Health Perspect 119:1604–1609
30. Kais Z, Parvin JD (2012) Centrosome regulation and breast cancer. In: Schatten H (ed) The centrosome, Chap 14. Springer Science and Business Media, New York
31. Kammerer S, Roth RB, Hoyal CR, Reneland R, Marnellos G, Kiechle M, Schwarz-Boeger U, Griffiths LR, Ebner F, Rehbock J, Cantor CR, Nelson MR, Brown A (2005) Association of the NuMA region on chromosome 11q13 with breast cancer susceptibility. Proc Natl Acad Sci U S A 102(6):2004–2009
32. Kasper M, Regl G, Frischauf AM, Aberger F (2006) GLI transcription factors: mediators of oncogenic hedgehog signalling. Eur J Cancer 42:437–445
33. Keri RA, Ho SM, Hunt PA, Knudsen KE, Soto AM et al (2007) An evaluation of evidence for the carcinogenic activity of bisphenol a. Reprod Toxicol 24:240–252
34. Korzeniewski N, Duensing S (2012) Disruption of centrosome duplication control and induction of mitotic instability by the high-risk human papillomavirus oncoproteins E6 and E7. In: Schatten H (ed) The centrosome, Chap 12. Springer Science and Business Media, New York
35. Korzeniewski N, Wheeler S, Chatterjee P et al (2010) A novel role of the aryl hydrocarbon receptor (AhR) in centrosome amplification – implications for chemoprevention. Mol Cancer 9:153

36. Krämer A, Anderhub S, Maier B (2012) Mechanisms and consequences of centrosome clustering in cancer cells. In: Schatten H (ed) The centrosome, Chap 17. Springer Science and Business Media, New York
37. Lang IA, Galloway TS, Scarlett A, Henley WE, Depledge M et al (2008) Association of urinary bisphenol a concentration with medical disorders and laboratory abnormalities in adults. JAMA 300:1303–1310
38. Leber B, Maier B, Fuchs F, Chi J, Riffel P, Anderhub S, Wagner L, Ho AD, Salisbury JL, Boutros M, Krämer A (2010) Proteins required for centrosome clustering in cancer cells. Sci Transl Med 2(33 33ra38):1–11
39. Levy YY, Lai EY, Remillard SP, Heintzelman MB, Fulton C (1996) Centrin is a conserved protein that forms diverse associations with centrioles and MTOCs in Naegleria and other organisms. Cell Motil Cytoskeleton 33:298–323
40. Li Y, Hu J (2015) Small GTPases act as cellular switches in the context of cilia. In: Schatten H (ed) The cytoskeleton in health and disease. Springer Science and Business Media, New York
41. Li Y, Lu W, Chen D, Boohaker RJ, Zhai L, Padmalayam I, Wennerberg K, Xu B, Zhang W (2015) KIFC1 is a novel potential therapeutic target for breast cancer. Cancer Biol Ther 16:1316–1322
42. Ling H, Peng L, Seto E, Fukasawa K (2012) Suppression of centrosome duplication and amplification by deacetylases. Cell Cycle 11:3779–3791
43. Lingle WL, Salisbury JL (1999) Altered centrosome structure is associated with abnormal mitoses in human breast tumors. Am J Pathol 155:1941–1951
44. Lingle WL, Salisbury JL (2000) The role of the centrosome in the development of malignant tumors. Curr Top Dev Biol 49:313–329
45. Lingle WL, Lutz WH, Ingle JN, Maihle NJ, Salisbury JL (1998) Centrosome hypertrophy in human breast tumors: implications for genomic stability and cell polarity. Proc Natl Acad Sci U S A 95:2950–2955
46. Liu Q, Zhang Y, Jernigan D, Fatatis A (2011) Survival and growth of prostate Cancer cells in the bone: role of the alpha-receptor for platelet-derived growth factor in supporting early metastatic foci. In: Fatatis A (ed) Signaling pathways and molecular mediators in metastasis. Springer, Dordrecht
47. Lutz W, Lingle WL, McCormick D, Greenwood TM, Salisbury JL (2001) Phosphorylation of centrin during the cell cycle and its role in centriole separation preceding centrosome duplication. J Biol Chem 276:20774–20780
48. Manandhar G, Schatten H, Sutovsky P (2005) Centrosome reduction during gametogenesis and its significance. Biol Reprod 72:2–13
49. Marchetti F, Mailhes JB, Bairnsfather L, Nandy I, London SN (1996) Dose-response study and threshold estimation of griseofulvin induced aneuploidy during female mouse meiosis I and II. Mutagenesis 11:195–200
50. Merdes A, Cleveland DA (1998) The role of NuMA in the interphase nucleus. J Cell Sci 111:71–79
51. Miao Y-L, Kikuchi K, Sun Q-Y, Schatten H (2009a) Oocyte aging: cellular and molecular changes, developmental potential and reversal possibility. Human Reprod Update 15(5):573–585
52. Miao Y-L, Sun Q-Y, Zhang X, Zhao J-G, Zhao M-T, Spate L, Prather RS, Schatten H (2009b) Centrosome abnormalities during porcine oocyte aging. Environ Mol Mutagen 50(8):666–671
53. Michaud EJ, Yoder BK (2006) The primary cilium in cell signaling and cancer. Cancer Res 66:6463–6467
54. Mittal K, Choi DH, Klimov S, Pawar S, Kaur R, Mitra AK, Gupta MV, Sams R, Cantuaria G, Rida PCG, Aneja R (2016) A centrosome clustering protein, KIFC1, predicts aggressive disease course in serous ovarian adenocarcinomas. J Ovarian Res 9(17):1–11
55. Mogensen MM, Malik A, Piel M, Bouckson-Castaing V, Bornens M (2000) Microtubule minus-end anchorage at centrosomal and non-centrosomal sites: the role of ninein. J Cell Sci 113:3013–3023

56. Olivero OA (2012) Centrosomal amplification and related abnormalities induced by nucleoside analogs. In: Schatten H (ed) The centrosome, Chap 16. Springer Science and Business Media, New York
57. Pacchierotti F, Ranaldi R, Eichenlaub-Ritter U, Attia S, Adler ID (2008) Evaluation of aneugenic effects of bisphenol a in somatic and germ cells of the mouse. Mutat Res 651(1–2):64–70
58. Pan J, Snell W (2007) The primary cilium: keeper of the key to cell division. Cell 129:1255–1257
59. Panda D, Rathinasamy K, Santra MK, Wilson L (2005) Kinetic suppression of microtubule dynamic instability by griseofulvin: implications for its possible use in the treatment of cancer. Proc Natl Acad Sci U S A 102:9878–9883
60. Prins GS, Ye SH, Birch L, Ho SM, Kannan K (2011) Serum bisphenol a pharmacokinetics and prostate neoplastic responses following oral and subcutaneous exposures in neonatal Sprague-Dawley rats. Reprod Toxicol 31:1–9
61. Prosser SL, Fry AM (2012) Regulation of the centrosome cycle by protein degradation. In: Schatten H (ed) The centrosome, Chap 9. Springer Science and Business Media, New York
62. Quarmby LM, Parker JDK (2005) Cilia and the cell cycle? J Cell Biol 169(5):707–710
63. Rebacz B, Larsen TO, Clausen MH, Ronnest MH, Loffler H, Ho AD, Krämer A (2007) Identification of griseofulvin as an inhibitor of centrosomal clustering in a phenotype-based screen. Cancer Res 67:6342–6350
64. Saladino C, Bourke E, Morrison CG (2012) Centrosomes, DNA damage and aneuploidy. In: Schatten H (ed) The centrosome, Chap 13. Springer Science and Business Media, New York
65. Salisbury JL (1995) Centrin, centrosomes, and mitotic spindle poles. Curr Opin Cell Biol 7:39–45
66. Salisbury JL (2004) Centrosomes: Sfi1p and centrin unravel a structural riddle. Curr Biol 14:R27–R29
67. Salisbury JL, Suino KM, Busby R, Springett M (2002) Centrin-2 is required for centriole duplication in mammalian cells. Curr Biol 12:1287–1292
68. Saredi A, Howard L, Compton DA (1997) Phosphorylation regulates the assembly of NuMA in a mammalian mitotic extract. J Cell Sci 110:1287–1297
69. Satir P, Christensen ST (2008) Structure and function of mammalian cilia. Histochem Cell Biol 129:687–693
70. Schatten H (1977) Untersuchungen über die Wirkung von Griseofulvin in Seeigeleiern und in Mammalierzellen. Universität Heidelberg; 1977 (Effects of griseofulvin on sea urchin eggs and on mammalian cells. University of Heidelberg)
71. Schatten H (2008) The mammalian centrosome and its functional significance. Histochem Cell Biol 129:667–686
72. Schatten H (2013) Chapter 12: The impact of centrosome abnormalities on breast Cancer development and progression with a focus on targeting centrosomes for breast Cancer therapy. In: Schatten H (ed) Cell and molecular biology of breast Cancer. Springer Science and Business Media, LLC, Ney York
73. Schatten H (2014) Chapter 12: The role of centrosomes in cancer stem cell functions. In: Schatten H (ed) Cell and molecular biology and imaging of stem cells, 1st edn. Wiley, Hoboken, pp 259–279
74. Schatten H, Sun QY (2009) The functional significance of centrosomes in mammalian meiosis, fertilization, development, nuclear transfer, and stem cell differentiation. Environ Mol Mutagen 50(8):620–636
75. Schatten H, Sun Q-Y (2011) The significant role of centrosomes in stem cell division and differentiation. Microsc Microanal 17(4):506–512 Epub 2011 Jul 11
76. Schatten H, Sun Q-Y (2015a) Centrosome and microtubule functions and dysfunctions in meiosis: implications for age-related infertility and developmental disorders. Reprod Fertil Dev. https://doi.org/10.1071/RD14493 [Epub ahead of print]. PMID: 25903261
77. Schatten H, Sun Q-Y (2015b) Centrosome-microtubule interactions in health, disease, and disorders. In: Schatten H (ed) The cytoskeleton in health and disease. Springer, Science+Business Media New York

78. Schatten G, Schatten H, Bestor T, Balczon R (1982a) Taxol inhibits the nuclear movements during fertilization and induces asters in unfertilized sea urchin eggs. J Cell Biol 94:455–465

79. Schatten H, Schatten G, Petzelt C, Mazia D (1982b) Effects of griseofulvin on fertilization and early development of sea urchins. Independence of DNA synthesis, chromosome condensation, and cytokinesis cycles from microtubule-mediated events. Eur J Cell Biol 27:74–87

80. Schatten H, Wiedemeier A, Taylor M, Lubahn D, Greenberg NM, Besch-Williford C, Rosenfeld C, Day K, Ripple M (2000) Centrosomes-centriole abnormalities are markers for abnormal cell divisions and cancer in the transgenic adenocarcinoma mouse prostate (TRAMP) model. Biol Cell 92:331–340

81. Schiff PB, Fant J, Horwitz SB (1979) Promotion of microtubule assembly in vitro by taxol. Nature 277:665–667

82. Schneider L, Clement CA, Teilmann SC et al (2005) PDGFR alpha signaling is regulated through the primary cilium in fibroblasts. Curr Biol 15:1861–1866

83. Schoffski P (2009) Polo-like kinase (PLK) inhibitors in preclinical and early clinical development in oncology. Oncologist 14:559–570

84. Sharma N, Berbari NF, Yoder BK (2008) Ciliary dysfunction in developmental abnormalities and diseases. Curr Top Dev Biol 85:371–427

85. Ślusarz A, Shenouda NS, Sakla MS, Drenkhahn SK, Narula AS, MacDonald RS, Besch-Williford CL, Lubahn DB (2010) Common botanical compounds inhibit the hedgehog signaling pathway in prostate Cancer. Cancer Res 70(8):3382–3390

86. Sun QY, Schatten H (2006) Multiple roles of NuMA in vertebrate cells: review of an intriguing multifunctional protein. Front Biosci 11:1137–1146

87. Sun Q-Y, Schatten H (2007) Centrosome inheritance after fertilization and nuclear transfer in mammals. In: Sutovsky P (ed) Somatic cell nuclear transfer, Landes bioscience. Adv Exp Med Biol 591:58–71

88. Tang WY, Morey LM, Cheung YY, Birch L, Prins GS et al (2012) Neonatal exposure to estradiol/bisphenol a alters promoter methylation and expression of Nsbp1 and Hpcal1 genes and transcriptional programs of Dnmt3a/b and Mbd2/4 in the rat prostate gland throughout life. Endocrinology 153:42–55

89. Tarapore P, Ying J, Ouyang B, Burke B, Bracken B, Ho S-M (2014) Exposure to Bisphenol a correlates with early-onset prostate Cancer and promotes centrosome amplification and anchorage-independent growth in vitro. PLoS One 9(3):e90332 https://doi.org/10.1371/journal.pone.0090332

90. Veland IR, Awan A, Pedersen LB, Yoder BK, Christensen ST (2009) Primary cilia and signaling pathways in mammalian development, health and disease. Nephron Physiol 111:39–53

91. Wehland J, Herzog W, Weber K (1977) Interaction of griseofulvin with microtubules, microtubule protein and tubulin. J Mol Biol 111:329–342

92. Wheatley DN, Wang AM, Strugnell GE (1996) Expression of primary cilia in mammalian cells. Cell Biol Int 20:73–81

93. Xiao Y-X, Yang W-X (2016) KIFC1: a promising chemotherapy target for cancer treatment? Oncotarget 7(30):48656–48670

94. Yan B, Chng W-J (2012) The role of centrosomes in multiple myeloma. In: Schatten H (ed) The centrosome, Chap 15. Springer Science and Business Media, New York

95. Young A, Dictenberg JB, Purohit A, Tuft R, Doxsey SJ (2000) Cytoplasmic dynein-mediated assembly of pericentrin and γ tubulin onto centrosomes. Mol Biol Cell 11:2047–2056

Chapter 5
MicroRNAs as Regulators of Prostate Cancer Metastasis

Divya Bhagirath, Thao Ly Yang, Rajvir Dahiya, and Sharanjot Saini

Abstract Prostate cancer causes significant morbidity in men and metastatic disease is a major cause of cancer related deaths. Prostate metastasis is controlled by various cellular intrinsic and extrinsic factors, which are often under the regulatory control of various metastasis-associated genes. Given the dynamic nature of metastatic cancer cells, the various factors controlling this process are themselves regulated by microRNAs which are small non-coding RNAs. Significant research work has shown differential microRNA expression in primary and metastatic prostate cancer suggesting their importance in prostate pathogenesis. We will review the roles of different microRNAs in controlling the various steps in prostate metastasis.

Keywords Metastasis · Prostate cancer · microRNAs · ECM · EMT

5.1 Introduction

Prostate cancer (PCa) is the second most commonly diagnosed cancer among men in the United States and is the third leading cause of cancer related deaths. Metastatic disease accounts for ~16.5% of deaths from PCa [76]. Despite improvement in early screening methods and development of effective therapies, the rates at which aggressive prostate cancers are diagnosed are showing an increasing trend. Recent data analysis from patients with PCa within the United States has shown an increased rate of metastatic disease particularly in men of age group 55–69 years [21]. The rate of metastasis incidence has significantly increased at 2.74% per year from 2012 for all recorded cases. Moreover, there has been a steady increase in the incidence of metastatic PCa among white men as opposed to other races. Furthermore, these rates are expected to increase at 0.38% per year accounting for almost 42% of

D. Bhagirath · T. L. Yang · R. Dahiya · S. Saini (✉)
Department of Urology, Veterans Affairs Medical Center, San Francisco and University of California San Francisco, California, USA
e-mail: Sharanjot.Saini@ucsf.edu

© Springer International Publishing AG, part of Springer Nature 2018   83
H. Schatten (ed.), *Cell & Molecular Biology of Prostate Cancer*,
Advances in Experimental Medicine and Biology 1095,
https://doi.org/10.1007/978-3-319-95693-0_5

metastatic PCa cases by 2025 [51], which is an alarming increase for the most prevalent male cancer.

Tumor metastasis is a multistep process that involves dissemination of cancer cells from the primary site, their survival in the circulatory system, extravasation to the metastatic sites and subsequent colonization [14, 38]. In addition to their own genetic susceptibilities, cancer cells disseminated from primary sites depend on several growth regulatory signals such as those from chemokines and cytokines in the metastatic niche in order to survive and proliferate at secondary locations [14]. Given the inherent heterogeneity in primary tumors, one can expect emergence of clones that are fit to survive the adversities encountered during the entire process of metastasis. These tumor cells evolve both genetically and epigenetically to surmount the barriers of survival outside their primary niche [38]. There is activation of genes that facilitate metastatic progression, including those that control the epithelial-to-mesenchymal (EMT) pathways, invasion-migratory pathways and allow proteolysis and degradation of the extracellular matrix (ECM). TWIST1, SNAI1, SNAI2 and ZEB1 are some of the important transcription factors that regulate this important step of EMT activation [62]. These essential initiators of metastasis are often regulated in most tumors including PCa by small specialized RNAs called microRNAs [36]. *ANGPTL4, MMP1, PTGS2, EREG* are some of the genes that allow invasion and survival in the circulation and *IL11, IL6, PTHRP* are genes that facilitate subsequent colonization [62]. Further, these cells cooperate with the microenvironment and its constituent cell types such as fibroblasts and immune cells so as to attain aggressive phenotypes that entail metastatic progression [38]. MicroRNAs (miRNAs) are small non-coding regulatory RNAs which are often deregulated in tumors. In this book chapter, we will provide an overview on PCa metastasis following which we will review the roles of different miRNAs and their contribution towards different steps in PCa metastasis.

5.2 Prostate Cancer Metastasis

Prostate tumors are clinically defined as either indolent or aggressive. Majority of these tumors are localized and treated according to their stage or Gleason score which is determined primarily by biopsy sampling. Most tumors that are identified as aggressive or advanced at the time of first diagnosis and are often accompanied by micrometastasis at secondary locations. Of the different sites in the body, the skeletal system has especially high propensity to develop metastatic lesions in patients with prostate cancer [14, 30, 60]. The presence of bone metastasis in PCa patients in addition to visceral metastasis is associated with a significantly lower overall survival of 14 months [30]. Thus, signifying the importance of metastatic locations in determining disease prognosis. There are different hypotheses to explain the genetics of metastatic tumors. Primary prostate tumors are heterogeneous and are often comprised of different clonal populations that may give rise to metastasis [70]. Liu *et al* studied the association of primary tumors with their corresponding metastatic tumors and suggested a monoclonal origin for metastatic tumors. They analyzed the copy number alterations and genome wide nucleotide polymorphisms across tumors derived from different metastatic locations in the same patients and observed a genomically stable pattern among these different tumors [57]. However, a more recent study from Hong *et al.* suggests that metastatic tumors are different from primary tumors and that acquisition of favorable mutation such as those in tumor suppressor *TP53* contributes to metastatic tumor heterogeneity and confers metastatic potential to these cells. These mutations or variation in metastatic tumors may arise in order to provide better survival advantage against therapeutic interventions [41, 70]. Furthermore, they also suggest a cross-metastatic seeding pattern in patients where metastatic tumor subclones themselves seed new tumors in more secondary locations [41].

5.2.1   Factors Regulating PCa Metastasis

Several cell intrinsic and extrinsic factors play an important role in execution of a successful metastatic process. While factors intrinsic to a cell refer to the nature of cell type and its genetic composition, microenvironmental cues correspond to the extrinsic factors controlling metastasis. Tumors cells with extensive genomic instability are considered to be the potential candidates for metastatic initiation [62]. Prostate tumor cells are well known to harbor copy number alterations, chromosomal deletions, DNA rearrangements such as chromoplexy, chromothripsis, gene fusions and certain mutations [2, 3, 85]. Presence of *TMPRSS-ERG* gene fusions is a common event among prostate cancer cells. Fluorescence *in situ* hybridization in primary and corresponding metastatic tumors derived from patients undergoing radical prostectomies have shown association of *TMPRSS2-ERG* fusion among the tumors suggesting a common origin [35]. More recently, it has been shown that over-expression of *TMPRSS2 ERG* fusions in metastatic PCa cell line PC3M-Luc leads to increased bone metastasis in mice [25]. Epigenetic modifications also play an important role in imparting plasticity to the cell that facilitates initiation of metastatic events. EZH2, a histone lysine methyltransferase enzyme is over-expressed in prostate tumors from advanced stage prostate cancer patients and its expression directly correlates with metastatic progression of the disease [89]. Germline mutations in DNA repair genes such as *BRCA1, BRCA2, CHEK2, ATM, RAD51D* and *PALB2* is a significant prognostic factor for development of metastatic prostate cancer [66]. Stankiewicz *et al* recently identified *FBXL4* gene which is located on chromosome 6q to be deleted in metastatic bone tumors as well as the primary tumors suggesting its role as a possible tumor suppressor that is lost during metastatic progression [78]. Deletions/additions in parts of chromosomes is a commonly observed phenomenon in PCa tumors [70]. Chromosomal gain (1q, 3q, 7q, 8q, 17q and Xq) and loss (1q, 6q, 8p, 10q, 13q and 16q) are frequently observed chromosomal alterations in the prostate tumor genome [7]. Homeobox tumor suppressor gene *NKX3.1* that is involved in differentiation of the prostate gland is located at chromosome 8p and its loss has been linked to disease progression [4, 6, 40]. Many tumor-suppressor miRNA genes are located on these chromosomes. Studies from our lab have shown that chromosome 8p is frequently lost during PCa progression, harbors miRNAs that have tumor-suppressor functions and plays an essential role in regressing EMT transition in tumor cell [8–10].

5.2.2 Microenvironment and PCa Metastasis

The extracellular microenvironment is another important determinant in the process of metastatic cancer progression. The extracellular matrix (ECM) is formed of diverse matrix proteins including laminin, fibronectin, vitronectin, collagen, osteopontin and others in which stromal cells such as adipocytes, fibroblasts, immune cells and many tissue-specific cells are embedded. Together with blood and lymphatic vessels, they form a niche necessary for sustenance and proliferation of tumor cells. The stromal compartment itself is a strong prognostic indicator of progressing disease. Gene expression analysis of the stromal and epithelial compartment of tumor tissue has revealed a distinct stroma signature that significantly varies with a patient's Gleason score, thereby serving as an independent indicator of high grade PCa [87].

Expression of many basement membrane and matrix proteins are altered during PCa progression [79]. In addition, PCa cells also demonstrate altered patterns of cell adhesion molecules (CAMs) including cadherins and integrins that allow interaction with the surrounding matrix and facilitate cancer progression. The expression of integrins varies with the metastatic stage that in turn, allows protection and easier passage of tumor cells from the external barriers in the primary site, blood vessels and metastatic locations [79, 84]. Signaling through these surface molecules mediated by their interaction with ECM proteins or endothelial cells control motility, growth and proliferation of the disseminated cancer cell [84]. Proteases such as matrix metalloproteases (MMPs), urokinase plasminogen activator (uPA), and cathepsins are upregulated in the invasive disease, thus allowing for degradation of the basement membrane and dissemination of cancer cells through the matrix [79].

Soluble growth promoting signaling molecules such as growth factors, chemokines and chemo-attractants are other important players in mediating successful colonization at metastatic sites. As mentioned earlier, most PCa patients often develop bone metastasis [30]. Stephan Paget's "seed and soil" hypothesis and James Ewing's circulatory connection theory help explain this preferential metastatic spread of prostate tumor cells to bone [14, 84]. Insulin-like growth factor 1 (IGF-1), transforming Growth Factor- β (TGF-β), uPA, bone morphogenetic proteins (BMPs) and parathyroid hormone-related protein (PTHrP) are some of the soluble factors that facilitate the growth and survival of tumor cells in the bone microenvironment [60, 84]. Runt-related transcription factor (Runx2) is found to be overexpressed in metastatic PCa cell lines and has been shown to mediate bone-specific metastatic behavior in tumor cells [1]. In addition to these biochemical mediators, physical forces from the microenvironment also influence metastatic progression [45]. Pressure changes in the bone microenvironment have been shown to enhance the migratory behavior of PCa cell lines via induction of Chemokine (C-C motif) ligand 5 (CCL5) from the osteocytes [77].

5.2.3 Exosomes in Metastasis

Extracellular vesicles (EVs) are other important tumor cell messengers that facilitate cell-to-cell communication between primary tumor cells and distant metastatic locations. Exosomes are small EVs of size ranging from 30-100 nm that are secreted by several cell types [20]. These bilayer vesicles carry different biomolecules including: RNA, DNA, protein and lipids. A very recent study by Hoshino et al. have demonstrated that organ-specific metastatic patterns are governed by exosomes released by tumor cells. They injected tagged exosomes from breast and pancreatic cancer cells that specifically metastasize to lungs or bone to study patterns of their distribution *in vivo*. They observed organ-specific deposition of the exosomes that allows for colonization of the disseminated or injected tumor cells to the exact location of exosome deposition. This study suggests that exosomes functionally educate the metastatic locations beforehand to facilitate colonization by the tumor cells [42]. Further, these small vesicular particles are abundantly secreted in most biological fluids under normal and diseased conditions, thus profiling the exosomal particles offers a great opportunity to detect cancer and other pathological conditions at various stages of the disease. Specialized small RNAs such as microRNAs are often enriched in these EVs and are thought to mediate post-transcriptional gene controls in the receptor cells [88]. Thus, they not only function as biological mediators of metastatic progression but can also serve as markers for disease severity.

In conclusion, tumor metastasis is a dynamic process that is governed by several cell intrinsic and extrinsic factors and requires continuous changes in the tumor and surrounding cells for successful execution. Stable genetic changes in the tumors confer genetic fitness to initiate this process, however different steps in the process often require post-transcriptional regulation of several genes controlling the EMT pathways, proteolysis and secretion of factors that allow for cell survival and subsequent colonization. Small non-coding RNA molecules such as miRNAs are important mediators of such posttranscriptional gene controls. We will discuss in detail in the following sections how these specialized small RNAs influence prostate tumorigenesis with a focus on their roles in PCa metastasis.

5.3   MicroRNAs: Small Non–Coding Gene Regulators

MicroRNAs are 22–23 nt long, single stranded non-coding RNA molecules that play an important role in regulating gene transcription post-transcriptionally. They were first discovered in *Caenorhabditis elegans* in 1993 and since then many regulatory miRNAs have been identified. Several studies have demonstrated their role in tumor pathogenesis by regulation of cell growth, proliferation, apoptosis, cell cycle and cell differentiation [26]. Almost 60% of the human genome is regulated by these small RNAs [12]. They are synthesized as a long chain of polycistronic RNA that is cleaved by ribonucleases DROSHA and DICER into 22–23 nt long single stranded RNAs. They bind through complete or partial complementarity to the 3'-UTRs of target mRNAs and together with Argonaute protein form a RISC complex at the mRNA strand that catalyzes its degradation [26]. Their expression is often deregulated in cancer, where miRNAs that lose their expression in tumors are called as "tumor-suppressors" and those that are up-regulated are called "oncomirs". Studies from our and many other laboratories have provided insights into miRNAs and the important role they play in prostate tumorigenesis. In this chapter, we will be highlighting their contribution in the process of PCa metastasis.

5.3.1   MicroRNAs in Prostate Cancer Metastasis

Tumor cells undergo many gene expression changes to acquire metastatic ability. Gene expression analysis of metastatic tumors have revealed molecular signatures corresponding to cell cycle, transcription factors, signal transduction pathways that can be therapeutically targeted and can predict prognosis for disease recurrence [34, 52]. miRNAs also share this deregulatory behavior in metastatic tumors. Next generation sequencing of metastatic tumors have shown that a large number of small RNAs are differentially expressed and are believe to be important players in the metastatic process [71, 86, 90]. More recently, Xue *et al.* designed a computer algorithm to analyze different PCa metastasis datasets. Based on their observations they identified transcriptional factors AR, HOX6 and NKX2–2 that were altered in metastatic tumors and are believed to regulate the expression of various metastasis-related miRNAs. These TFs were then validated *in vitro* for their functional significance in controlling metastasis by miRNA regulation from prostate epithelial cell line RWPE1 [92]. Deregulation of miRNA expression is often accompanied by metastatic disease and miRNAs are essentially required for the process. We will be discussing roles of various miRNAs in different metastatic steps including: acquisition of EMT, regulation of factors responsible for tumor cell metastasis/colonization and regulation of the microenvironment to facilitate tumor progression.

5.3.1.1 miRNAs in Regulating Acquisition of EMT

Dissemination from the primary site is the first step in metastasis, which requires a tumor cell to undergo transition from an epithelial to a mesenchymal state. Epidermal Growth Factor Receptor *(EGFR)* is well known oncogene that is often overexpressed in many tumors and its expression is thought to promote bone metastasis in breast and prostate cancer. A very recent study by Day *et al.* has shown that circulating tumor cells in PCa patients with bone metastasis overexpress *EGFR* while tumor cells depend on *Her2* overexpression for their growth in the bone microenvironment [24]. Additionally, EGFR has been also shown to regulate the expression of miR-1 in PCa cell lines which in turn, controls EMT transcription factor TWIST1. Together, they form a mechanistic loop where EGFR expression increases with progression of disease and miR-1 expression decreases that leading to an increase in TWIST1, thus facilitating EMT progression [15]. It has been previously shown that miR-1 is implicated in metastatic disease and its expression is lost both in primary and metastatic tumors. It also functions as a tumor–suppressor that regulates genes related to the cell-cycle, apoptosis, DNA damage and inhibits the invasive phenotype of PCa cells lines [44]. miR-143 and miR145 are downregulated in bone metastatic tumors and their overexpression in PCa cell lines decreases invasion-migration and bone metastasis forming ability of cancer cell lines *in vi*vo [65]. miR-145, which is negatively correlated with HEF1 expression directly targets the 3'-UTR of *HEF1* mRNA and ablates its EMT conferring properties in PCa cell lines [37]. miR-29b when over-expressed in PC3 cancer cells reduces the invasiveness and lung and liver metastasis forming ability *in vivo*. Furthermore, it leads to up-regulation of E-Cadherin and down-regulation of EMT markers TWIST1, N-Cadherin and SNAI1 [69]. miR-182 and miR-203 target SNAI2 and induce epithelial phenotypes and self-sufficient growth ability in prostate cells EPT1 [68]. Human enhancer of filamentation-1 (HEF-1) is highly expressed in bone metastatic specimens from PCa patients and it regulates EMT and aggressiveness in PC3 cells. miR-409-3p/5p are members of the delta like 1 homolog-deiodinase, iodothyronine 3 (DLK1-DIO3) cluster which are known to be involved in prostate metastasis [47]. miR-409-3p and 5p have been shown to be overexpressed in PCa patient serum and are involved in mediating tumorigenicity, EMT and bone metastasis *in vivo* in PCa cell lines upon its overexpression [47]. miR-154 and miR-379 are other miRNAs found in the same DLK1-DIO3 cluster that are overexpressed in metastatic bone lesions from PCa patients. Inhibition of these miRNAs in bone metastatic PCa cell lines leads to acquisition of MET and reduced bone and soft tissue metastasis from these cell lines [39]. Furthermore, combined overexpression of all 4 miRNAs found in the DLK1-DIO3 cluster including miR-409-3p/5p, miR-154 and miR-379 was shown to promote EMT in PCa cell lines. They are believed to function in activating oncogenic pathways including Ras, hypoxia-inducible factor (HIF), WNT and TGFβ signaling by targeting tumor-suppressors cohesion sub-unit SA-2 (STAG2), SMAD7, Von Hippel-Lindau tumor-suppressor (VHL) and polyhomeotic-like protein 3 (PHC3) [39]. miR-195 located on chromosome 17p13.1 functions as a tumor-suppressor and is found to be downregulated in high grade prostate tumors. Ribosomal protein

S6 kinase (RPS6KB-1) was found to be a miR-195 target gene. RPS6KB-1 knockdown restored cell migration, invasion and increased apoptosis observed in PCa cells as a result of miR-195 overexpression. Alterations in the miR-195-RPS6KB-1 axis were shown to regulate expression of MMP-9, BAD, E-Cadherin and VEGF that are involved in PCa progression. This study established the role of miR-195 in preventing PCa metastasis [11]. More recently, it has been shown that increased expression of miR-301a is associated with PCa recurrence in patients undergoing radical prostatectomy. Ectopic overexpression of miR-301a in PCa cell lines PC3 and LNCaP led to increased cell growth, invasion and migration. miR-301a directly targets p63, a member of the p53 family, that in turn alters the expression of EMT proteins E-Cadherin and transcription factor ZEB1 [61]. Recent studies from our lab have demonstrated that genomic loss of chromosome 8p21 is associated with PCa progression. We have identified that miRNAs, miR-3622a and miR-3622b, which are located within this genomic region play an important role in regulating PCa progression [8, 9]. We found that miR-3622a is widely downregulated in PCa and that miR-3622a represses PCa EMT by directly targeting *ZEB1* and *SNAI2*. miR-3622a loss allows tumor cells to acquire a mesenchymal phenotype, promoting invasion and metastasis [8]. Ectopic overexpression of miR-3622b in PCa cell lines led to reduced cellular viability, proliferation, invasiveness, migration and increased apoptosis. miR-3622b overexpression *in vivo* induced regression of established prostate tumor xenografts and miR-3622b was found to directly target EGFR [9].

5.3.1.2 Regulatory Effect of miRNAs on PCa Metastasis–Associated Signaling Pathways and Other Factors

miR-1 has been shown be downregulated in PCa tumors, more so in the metastatic samples. It directly targets SRC which is known to be a promoter of PCa metastasis [59]. miR-30 is commonly downregulated in PCa tumors and increases with SRC inhibitors. Overexpression of miR-30 in VCaP cells leads to reduction in expression of EMT genes and downregulation of Ets- related genes (ERG). Thus, miR-30 plays an important role in modulating the SRC/EGF and ERG pathways in tumor cells [50]. In Ras-activated xenograft tumors, miR-34a negatively regulates the expression of WNT signaling protein transcription factor 7 (TCF7) and anti-apoptotic protein baculoviral inhibitor of apoptosis repeat containing 5 (BIRC5), both of which are required for successful PCa metastasis in Ras-driven tumors [17]. Levels of miR-194 are often elevated in serum as well as tissues of PCa patients and can serve as a marker for disease recurrence. Forced overexpression in PCa cell lines have shown a pro-metastatic and invasive role for miR-194 in PCa tumorigenesis and it has been shown to directly target ubiquitin ligase suppressor of cytokine signaling 2 (SOCS2) protein. SOCS2 further regulates ubiquitination of two important kinases JAK and FLT3 that in turn, deregulates STAT3-mediated expression of pro-metastatic genes [23]. Loss of miR-15,16 and a concomitant increase in miR-21 activates TGFβ signaling pathways and plays an important role in bone colonization of PCa cells [5]. Studies from our group have identified many miRNAs that exert an anti-metastatic effect by targeting key metastatic genes. miR-203 is frequently lost in metastatic tumors and bone metastatic PCa cell lines. miR-203 over-expression in PCa cells alters the EMT markers, reduces invasion-migration and targets key metastatic genes including survivin, *ZEB2, SMAD4, DLX5* and *RUNX2* which is a known bone metastasis promoting transcription factor [73]. More recently, we identified miR-466 as another anti-metastasis miRNA that is downregulated in PCa tumors and its overexpression reduces PCa tumor and metastasis growth *in vivo*, targets *RUNX2* and alters the expression of *RUNX2* target genes including *MMP11, Angiopoietin (ANGPT1), ANGPT4, Osteopontin (OPN) and Osteocalcein (BGLAP)* [19].

Cancer stem cells (CSCs) are considered to be a precursor cell population for metastatic tumors and are defined by expression of cell surface markers such as CD44 [83]. These highly tumorigenic and metastatic cell populations have also been identified and characterized in PCa tumors [63, 64]. miR-34a is shown to be downregulated in CD44+ cell populations in PCa xenografts as well as purified CD44+ cells from PCa cell lines. Ectopic overexpression in PCa cell lines have shown that miR-34a directly represses CD44 expression and reduces the migratory and invasive phenotype of CD44+ cells, thus diminishing their metastatic potential [54]. Studies from our lab have demonstrated the role of miR-708 in reducing the tumorigenic potential of CD44+ PCa cells. It was shown to target CD44 and Ser/Thr kinase AKT2, thus altering tumor progression [72]. More recently, we identified that miR-383 located on chromosome 8p is lost during PCa progression and has an inhibitory effect on the CD44+ PCa cell population [10]. miR-128, miR-199-3p, miR-320 and miR141 are some of the other miRNAs that have been shown to regulate prostate metastasis by directly regulating the tumor-initiating stem populations in PCa [43, 46, 55, 56].

5.3.1.3 Microenvironmental Control of miRNAs in Regulation of PCa Metastasis

Metastasis is often marked by loss of BM protein that facilitates invasion of disseminated tumor cells. miR-205 plays an important role in deposition of the major BM protein laminin in prostate tissues. miR-205, along with TP63, regulates the deposition of BM protein and its expression is often lost with PCa tumor progression [31]. Expression of miR-25 is reduced in prostate cancer stem cells (PCSCs) when compared to differentiated luminal cells. Its overexpression has been shown to target expression of integrins αv and α6 in metastatic PCa cell lines and it leads to reduced migration and decreased metastasis *in vivo* [93]. miR-1207-3p is also lost during PCa progression. It has been shown to directly target fibronectin type III domain containing protein (FNDC1) that in turn, regulates fibronectin (FN1) and Androgen receptor (AR) in PCa cell lines. Loss of miR1207-3p is marked by increased expression of FNDC1/FN1/AR that is associated with PCa aggressiveness [22]. In order to understand the role of miRNAs with increasing Gleason grade, when tumors from Gleason grades 3, 4 and 5 were subjected to miRNA gene expression analysis, the results demonstrated miR-29c, miR-34a and mir-141 as differentially expressed miRs that had reduced expression with increasing grade. miRNAs function as tumor-suppressors and their overexpression reduces tumor cell migration and downregulation of ECM, focal adhesion kinase and MAPK13 pathways [53]. Syndecan-1 is another ECM protein that positively regulates levels of miR-331-3p which in turn, contributes towards increased EMT and aggressiveness in prostate tumors [29].

Tumor-associated stroma also undergoes changes in response to progressing tumors. Mesenchymal stem cells (MSCs) have been shown to migrate to tumors and promote prostate tumorigenicity [49]. Co-culture of MSCs with PCa cell lines *in vitro* have been shown to induce more adipogenic differentiation in these cells that is mediated through IL6. Expression of IL6 in MSCs is further regulated by let-7 miRNA, which is downregulated in tumor cells co-cultured with MSCs, thus signifying an important regulatory role of miRNA let-7 in determining the reactivity of tumor stroma [80]. Pre-adipocytes have been shown to be associated with the prostate tumor microenvironment as opposed to normal prostate tissues. They enhance the invasiveness and metastasis of PCa cell lines via upregulation of miR-301a which targets AR expression in tumors and in turn, affects expression of metastasis associated genes *MMP9, SMAD3* and TGF-β3 in PCa cells [91]. Cancer-associated fibroblasts (CAFs) are reactive fibroblasts that are often found in the tumor microenvironment. miRNA analysis of CAFs derived from patients with PCa tumors revealed miR-133a as highly expressed miRNA in these cells. miR-133a released from CAFs functions as a soluble paracrine factor that activates adjacent normal fibroblast to attain a reactive phenotype [27]. In the bone microenvironment, osteoblasts are the main effector cells that allow metastatic colonization by tumor cells under the influence of various factors. It has been shown that osteoblasts secrete Wnt1-induced secreted protein 1 (WISP1) that is released in conditioned media and acts on PCa cell lines to increase their invasion/migration abilities as well increasing the expression of vascular cell adhesion molecule 1 (VCAM1). This effect is mediated by miR-126 downregulation driven by osteoblast-derived WISP1 which further regulates αvβ1/p38 and FAK pathways in PCa cell lines [81]. miR-409-3p/5p are other miRNA that are found elevated in CAFs in prostate tumors. They are released by EVs from CAFs and upon their uptake by PCa cells mediate tumor cell EMT and aggressiveness [48].

Prolyl 4-hydroxylase alpha polypeptide 1 (P4HA1) enzyme is involved in proper folding of pro-collagen chains. It has been shown to be overexpressed during aggressive PCa and is regulated directly by miR-124. miR-124 is downregulated in high grade PCa and is transcriptionally regulated by EZH2 and transcriptional co-repressor C-terminal binding protein 1 (CtBP1) both of which are increased in aggressive PCa tumors [13].

Comparisons of prostate epithelial cell line P69 with respect to its metastatic subline M12 have shown altered miRNA expression among the two cell lines [16]. miR-130b is down-regulated in the metastatic M12 cell line, PC3, DU145 as well as in prostate tumors. It functions as a tumor suppressor and reduces invasion-migration ability of tumor cells. MMP2 is a direct target of miR-130b and exerts its invasive effect on metastasis as a result of miR-130b down-regulation [16]. miR-296-3p is over-expressed in M12, thereby down-regulating expression of intercellular cellular cell adhesion molecule 1 (ICAM1) which in turn, confers a protective effect to circulating tumor cells against natural killer (NK) cells [58].

miR-323 is upregulated in PCa cell lines and is shown to directly target adiponectin receptor 1 (AdipoR1) which in turn, negatively regulates vascular endothelial growth factor-A (VEGF-A)-mediated neovascularization. Thus, mir-323 mediated down-regulation of AdipoR1 facilitates formation of new blood vessels for the growing tumor [32]. Chemokine receptor CXCR4 is elevated in metastatic cell lines and is negatively regulated by miR-494-3p. Ectopic over-expression of miR-494-3p in PCa cell lines inhibits cell invasion and migration [75]. Circulating serum levels of miR-375 have been putatively linked to circulating tumor cells (CTCs) in metastatic CRPC patients. However, miR-375 have been shown to negatively regulate EMT and invasion in PCa cells and targets oncogene YAP1 which is often elevated in invasive PCa tumors. Furthermore, miR-375 is under negative regulation of ZEB1 which enables EMT in PCa cells, thus forming an axis of ZEB1, mir-375 and YAP1 that controls epithelial cell EMT and MET transitions [74].

5.4   Conclusions and Future Directions

Evidences from the literature suggest an essential role for miRNAs in the metastatic process (Fig. 5.1). In most PCa metastases, regulatory small RNAs are lost during tumor progression and function mostly as tumor suppressors. Thus, they inhibit

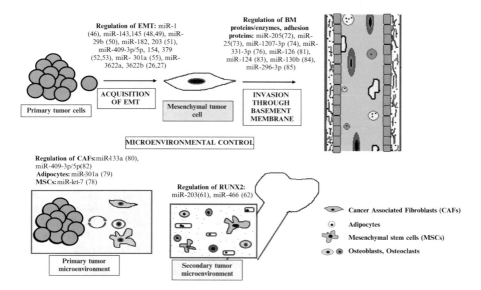

Fig. 5.1 Schematic representation of PCa metastasis and regulation of different steps by various miRNAs

metastatic initiation via EMT or regulate growth of primary tumor cells in the primary or metastatic microenvironment via control over important growth factors, chemokines, ECM and stromal components. In addition to their functional effects in mediating PCa metastasis, expression in tissues and circulatory are often important indicators of disease severity. Non-invasive sampling of cell free or EV-derived RNAs in the serum, plasma or urine offers a great opportunity for sensitive detection of metastatic disease [28]. Moreover, much research these days is focused on utilizing the therapeutic potential of miRNA in cancers. Different ways for delivering miRNA to their target cells i.e. via nanoparticles, liposomes, viral particle-mediated transfer or polyethylene glycol (PEG)-based particles, have been developed for systemic and local delivery of miRNAs [18, 66]. Atelocollagen particle-mediated delivery of miR-16 and chitosan nanoparticle-derived delivery of miR-34a to PCa xenografts *in vivo* have highlighted the promising effect of miRNA delivery in inhibiting advanced prostate cancer [33, 82]. Given the multifaceted roles of miRNAs, more research efforts are needed to improve PCa detection and the efficacy of disease therapeutics utilizing small regulatory miRNAs.

Acknowledgements We thank Dr. Roger Erickson for his support with preparation of the manuscript. Research in authors' lab is supported by *the National Cancer Institute at the NIH* (Grant Number RO1CA177984).

Conflict of Interest None.

References

1. Akech J, Wixted JJ, Bedard K, Van der Deen M, Hussain S, Guise TA, Van Wijnen AJ, Stein JL, Languino LR, Altieri DC, Pratap J, Keller E, Stein GS, Lian JB (2010) Runx2 association with progression of prostate cancer in patients: mechanisms mediating bone osteolysis and osteoblastic metastatic lesions. Oncogene 29:811–821
2. Baca SC, Prandi D, Lawrence MS, Mosquera JM, Romanel A, Drier Y, Park K, Kitabayashi N, Macdonald TY, Ghandi M, Van Allen E, Kryukov GV, Sboner A, Theurillat JP, Soong TD, Nickerson E, Auclair D, Tewari A, Beltran H, Onofrio RC, Boysen G, Guiducci C, Barbieri CE, Cibulskis K, Sivachenko A, Carter SL, Saksena G, Voet D, Ramos AH, Winckler W, Cipicchio M, Ardlie K, Kantoff PW, Berger MF, Gabriel SB, Golub TR, Meyerson M, Lander ES, Elemento O, Getz G, Demichelis F, Rubin MA, Garraway LA (2013) Punctuated evolution of prostate cancer genomes. Cell 153:666–677
3. Berger MF, Lawrence MS, Demichelis F, Drier Y, Cibulskis K, Sivachenko AY, Sboner A, Esgueva R, Pflueger D, Sougnez C, Onofrio R, Carter SL, Park K, Habegger L, Ambrogio L, Fennell T, Parkin M, Saksena G, Voet D, RamoS AH, Pugh TJ, Wilkinson J, Fisher S, Winckler W, Mahan S, Ardlie K, Baldwin J, Simons JW, Kitabayashi N, Macdonald TY, Kantoff PW, Chin L, Gabriel SB, Gerstein MB, Golub TR, Meyerson M, Tewari A, Lander ES, Getz G, Rubin MA, Garraway LA (2011) The genomic complexity of primary human prostate cancer. Nature 470:214–220
4. Bhatia-Gaur R, Donjacour AA, Sciavolino PJ, Kim M, Desai N, Young P, Norton CR, Gridley T, Cardiff RD, Cunha GR, Abate-Shen C, Shen MM (1999) Roles for Nkx3.1 in prostate development and cancer. Genes Dev 13:966–977

5. Bonci D, Coppola V, Patrizii M, Addario A, Cannistraci A, Francescangeli F, Pecci R, Muto G, Collura D, Bedini R, Zeuner A, Valtieri M, Sentinelli S, Benassi MS, GalluccI M, Carlini P, Piccolo S, de Maria R (2016) A microRNA code for prostate cancer metastasis. Oncogene 35:1180–1192

6. Bowen C, Bubendorf L, Voeller HJ, Slack R, Willi N, Sauter G, Gasser TC, Koivisto P, Lack EE, Kononen J, Kallioniemi OP, Gelmann EP (2000) Loss of NKX3.1 expression in human prostate cancers correlates with tumor progression. Cancer Res 60:6111–6115

7. Boyd LK, Mao X, Lu YJ (2012) The complexity of prostate cancer: genomic alterations and heterogeneity. Nat Rev Urol 9:652–664

8. Bucay N, Bhagirath D, Sekhon K, Yang T, Fukuhara S, Majid S, Shahryari V, Tabatabai Z, Greene KL, Hashimoto Y, Shiina M, Yamamura S, Tanaka Y, Deng G, Dahiya R, Saini S (2017) A novel microRNA regulator of prostate cancer epithelial-mesenchymal transition. Cell Death Differ 24:1263–1274

9. Bucay N, Sekhon K, Majid S, Yamamura S, Shahryari V, Tabatabai ZL, Greene K, Tanaka Y, Dahiya R, Deng G, Saini S (2016a) Novel tumor suppressor microRNA at frequently deleted chromosomal region 8p21 regulates epidermal growth factor receptor in prostate cancer. Oncotarget 7:70388–70403

10. Bucay N, Sekhon K, Yang T, Majid S, Shahryari V, Hsieh C, Mitsui Y, Deng G, Tabatabai ZL, Yamamura S, Calin GA, Dahiya R, Tanaka Y, Saini S (2016b) MicroRNA-383 located in frequently deleted chromosomal locus 8p22 regulates CD44 in prostate cancer. Oncogene

11. Cai C, Chen QB, Han ZD, Zhang YQ, He HC, Chen JH, Chen YR, Yang SB, Wu YD, Zeng YR, Qin GQ, Liang YX, Dai QS, Jiang FN, Wu SL, Zeng GH, Zhong WD, Wu CL (2015) miR-195 inhibits tumor progression by targeting RPS6KB1 in human prostate Cancer. Clin Cancer Res 21:4922–4934

12. Calin GA, Croce CM (2006) MicroRNA signatures in human cancers. Nat Rev Cancer 6:857–866

13. Chakravarthi BV, Pathi SS, Goswami MT, Cieslik M, Zheng H, Nallasivam S, Arekapudi SR, Jing X, Siddiqui J, Athanikar J, Carskadon SL, Lonigro RJ, Kunju LP, Chinnaiyan AM, Palanisamy N, Varambally S (2014) The miR-124-prolyl hydroxylase P4HA1-MMP1 axis plays a critical role in prostate cancer progression. Oncotarget 5:6654–6669

14. Chambers AF, Groom AC, Macdonald IC (2002) Dissemination and growth of cancer cells in metastatic sites. Nat Rev Cancer 2:563–572

15. Chang YS, Chen WY, Yin JJ, Sheppard-Tillman H, Huang J, Liu YN (2015) EGF receptor promotes prostate Cancer bone metastasis by downregulating miR-1 and activating TWIST1. Cancer Res 75:3077–3086

16. Chen Q, Zhao X, Zhang H, Yuan H, Zhu M, Sun Q, Lai X, Wang Y, Huang J, Yan J, Yu J (2015a) MiR-130b suppresses prostate cancer metastasis through down-regulation of MMP2. Mol Carcinog 54:1292–1300

17. Chen WY, Liu SY, Chang YS, Yin JJ, Yeh HL, Mouhieddine TH, Hadadeh O, Abou-Kheir W, Liu YN (2015b) MicroRNA-34a regulates WNT/TCF7 signaling and inhibits bone metastasis in Ras-activated prostate cancer. Oncotarget 6:441–457

18. Chen Y, Gao DY, Huang L (2015c) In vivo delivery of miRNAs for cancer therapy: challenges and strategies. Adv Drug Deliv Rev 81:128–141

19. Colden M, Dar AA, Saini S, Dahiya PV, Shahryari V, Yamamura S, Tanaka Y, Stein G, Dahiya R, Majid S (2017) MicroRNA-466 inhibits tumor growth and bone metastasis in prostate cancer by direct regulation of osteogenic transcription factor RUNX2. Cell Death Dis 8:e2572

20. Colombo M, Raposo G, Thery C (2014) Biogenesis, secretion, and intercellular interactions of exosomes and other extracellular vesicles. Annu Rev Cell Dev Biol 30:255–289

21. Dalela D, Sun M, Diaz M, Karabon P, Seisen T, Trinh QD, Menon M, Abdollah F (2017) Contemporary trends in the incidence of metastatic prostate Cancer among US men: Results from Nationwide analyses. Eur Urol Focus

22. Das DK, Naidoo M, Ilboudo A, Park JY, Ali T, Krampis K, Robinson BD, Osborne JR, Ogunwobi OO (2016) miR-1207-3p regulates the androgen receptor in prostate cancer via FNDC1/fibronectin. Exp Cell Res 348:190–200

23. Das R, Gregory PA, Fernandes RC, Denis I, Wang Q, Townley SL, Zhao SG, Hanson AR, Pickering MA, Armstrong HK, Lokman NA, Ebrahimie E, Davicioni E, Jenkins RB, Karnes RJ, Ross AE, Den RB, Klein EA, Chi KN, Ramshaw HS, Williams ED, Zoubeidi A, Goodall GJ, Feng FY, Butler LM, Tilley WD, Selth LA (2017) MicroRNA-194 promotes prostate Cancer metastasis by inhibiting SOCS2. Cancer Res 77:1021–1034

24. Day KC, Lorenzatti Hiles G, Kozminsky M, Dawsey SJ, Paul A, Broses LJ, Shah R, Kunja LP, Hall C, Palanisamy N, Daignault-Newton S, El-Sawy L, Wilson SJ, Chou A, Ignatoski KW, Keller E, Thomas D, Nagrath S, Morgan T, Day ML (2017) HER2 and EGFR overexpression support metastatic progression of prostate Cancer to bone. Cancer Res 77:74–85

25. Deplus R, Delliaux C, Marchand N, Flourens A, Vanpouille N, Leroy X, de Launoit Y, Duterque-Coquillaud M (2017) TMPRSS2-ERG fusion promotes prostate cancer metastases in bone. Oncotarget 8:11827–11840

26. di Leva G, Garofalo M, Croce CM (2014) MicroRNAs in cancer. Annu Rev Pathol 9:287–314

27. Doldi V, Callari M, Giannoni E, D'Aiuto F, Maffezzini M, Valdagni R, Chiarugi P, Gandellini P, Zaffaroni N (2015) Integrated gene and miRNA expression analysis of prostate cancer associated fibroblasts supports a prominent role for interleukin-6 in fibroblast activation. Oncotarget 6:31441–31460

28. Fendler A, Stephan C, Yousef GM, Kristiansen G, Jung K (2016) The translational potential of microRNAs as biofluid markers of urological tumours. Nat Rev Urol 13:734–752

29. Fujii T, Shimada K, Tatsumi Y, Tanaka N, Fujimoto K, Konishi N (2016) Syndecan-1 up-regulates microRNA-331-3p and mediates epithelial-to-mesenchymal transition in prostate cancer. *Mol Carcinog* 55:1378–1386

30. Gandaglia G, Karakiewicz PI, Briganti A, Passoni NM, Schiffmann J, Trudeau V, Graefen M, Montorsi F, Sun M (2015) Impact of the site of metastases on survival in patients with metastatic prostate Cancer. Eur Urol 68:325–334

31. Gandellini P, Profumo V, Casamichele A, Fenderico N, Borrelli S, Petrovich G, Santilli G, Callari M, Colecchia M, Pozzi S, de Cesare M, Folini M, Valdagni R, Mantovani R, Zaffaroni N (2012) miR-205 regulates basement membrane deposition in human prostate: implications for cancer development. Cell Death Differ 19:1750–1760

32. Gao Q, Yao X, Zheng J (2015) MiR-323 inhibits prostate Cancer vascularization through adiponectin receptor. Cell Physiol Biochem 36:1491–1498

33. Gaur S, Wen Y, Song JH, Parikh NU, Mangala LS, Blessing AM, Ivan C, Wu SY, Varkaris A, Shi Y, Lopez-Berestein G, Frigo DE, Sood AK, Gallick GE (2015) Chitosan nanoparticle-mediated delivery of miRNA-34a decreases prostate tumor growth in the bone and its expression induces non-canonical autophagy. Oncotarget 6:29161–29177

34. Glinsky GV, Glinskii AB, Stephenson AJ, Hoffman RM, Gerald WL (2004) Gene expression profiling predicts clinical outcome of prostate cancer. J Clin Invest 113:913–923

35. Guo CC, Wang Y, Xiao L, Troncoso P, Czerniak BA (2012) The relationship of TMPRSS2-ERG gene fusion between primary and metastatic prostate cancers. Hum Pathol 43:644–649

36. Guo F, Parker Kerrigan BC, Yang D, Hu L, Shmulevich I, Sood AK, Xue F, Zhang W (2014) Post-transcriptional regulatory network of epithelial-to-mesenchymal and mesenchymal-to-epithelial transitions. J Hematol Oncol 7:19

37. Guo W, Ren D, Chen X, Tu X, Huang S, Wang M, Song L, Zou X, Peng X (2013) HEF1 promotes epithelial mesenchymal transition and bone invasion in prostate cancer under the regulation of microRNA-145. J Cell Biochem 114:1606–1615

38. Gupta GP, Massague J (2006) Cancer metastasis: building a framework. Cell 127:679–695

39. Gururajan M, Josson S, Chu GC, Lu CL, Lu YT, Haga CL, Zhau HE, Liu C, Lichterman J, Duan P, Posadas EM, Chung LW (2014) miR-154* and miR-379 in the DLK1-DIO3 microRNA mega-cluster regulate epithelial to mesenchymal transition and bone metastasis of prostate cancer. Clin Cancer Res 20:6559–6569

40. He WW, Sciavolino PJ, Wing J, Augustus M, Hudson P, Meissner PS, Curtis RT, Shell BK, Bostwick DG, Tindall DJ, Gelmann EP, Abate-Shen C, Carter KC (1997) A novel human prostate-specific, androgen-regulated homeobox gene (NKX3.1) that maps to 8p21, a region frequently deleted in prostate cancer. Genomics 43:69–77

41. Hong MK, Macintyre G, Wedge DC, Van Loo P, Patel K, Lunke S, Alexandrov LB, Sloggett C, Cmero M, Marass F, Tsui D, Mangiola S, Lonie A, Naeem H, Sapre N, Phal PM, Kurganovs N, Chin X, Kerger M, Warren AY, Neal D, Gnanapragasam V, Rosenfeld N, Pedersen JS, Ryan A, Haviv I, Costello AJ, Corcoran NM, Hovens CM (2015) Tracking the origins and drivers of subclonal metastatic expansion in prostate cancer. Nat Commun 6:6605

42. Hoshino A, Costa-Silva B, Shen TL, Rodrigues G, Hashimoto A, Tesic Mark M, Molina H, Kohsaka S, di Giannatale A, Ceder S, Singh S, Williams C, Soplop N, Uryu K, Pharmer L, King T, Bojmar L, Davies AE, Ararso Y, Zhang T, Zhang H, Hernandez J, Weiss JM, Dumont-Cole VD, Kramer K, Wexler LH, Narendran A, Schwartz GK, Healey JH, Sandstrom P, Labori KJ, Kure EH, Grandgenett PM, Hollingsworth MA, de Sousa M, Kaur S, Jain M, Mallya K, Batra SK, Jarnagin WR, Brady MS, Fodstad O, Muller V, Pantel K, Minn AJ, Bissell MJ, Garcia BA, Kang Y, Rajasekhar VK, Ghajar CM, Matei I, Peinado H, Bromberg J, Lyden D (2015) Tumour exosome integrins determine organotropic metastasis. Nature 527:329–335

43. Hsieh IS, Chang KC, Tsai YT, Ke JY, Lu PJ, Lee KH, Yeh SD, Hong TM, Chen YL (2013) MicroRNA-320 suppresses the stem cell-like characteristics of prostate cancer cells by down-regulating the Wnt/beta-catenin signaling pathway. Carcinogenesis 34:530–538

44. Hudson RS, Yi M, Esposito D, Watkins SK, Hurwitz AA, Yfantis HG, Lee DH, Borin JF, Naslund MJ, Alexander RB, Dorsey TH, Stephens RM, Croce CM, Ambs S (2012) MicroRNA-1 is a candidate tumor suppressor and prognostic marker in human prostate cancer. Nucleic Acids Res 40:3689–3703

45. Jaalouk DE, Lammerding J (2009) Mechanotransduction gone awry. Nat Rev Mol Cell Biol 10:63–73

46. Jin M, Zhang T, Liu C, Badeaux MA, Liu B, Liu R, Jeter C, Chen X, Vlassov AV, Tang DG (2014) miRNA-128 suppresses prostate cancer by inhibiting BMI-1 to inhibit tumor-initiating cells. Cancer Res 74:4183–4195

47. Josson S, Gururajan M, Hu P, Shao C, Chu GY, Zhau HE, Liu C, Lao K, Lu CL, Lu YT, Lichterman J, Nandana S, LI Q, Rogatko A, Berel D, Posadas EM, Fazli L, Sareen D, Chung LW (2014) miR-409-3p/−5p promotes tumorigenesis, epithelial-to-mesenchymal transition, and bone metastasis of human prostate cancer. Clin Cancer Res 20:4636–4646

48. Josson S, Gururajan M, Sung SY, Hu P, Shao C, Zhau HE, Liu C, Lichterman J, Duan P, Li Q, Rogatko A, Posadas EM, Haga CL, Chung LW (2015) Stromal fibroblast-derived miR-409 promotes epithelial-to-mesenchymal transition and prostate tumorigenesis. Oncogene 34:2690–2699

49. Jung Y, Kim JK, Shiozawa Y, Wang J, Mishra A, Joseph J, Berry JE, Mcgee S, Lee E, Sun H, Wang J, Jin T, Zhang H, Dai J, Krebsbach PH, Keller ET, Pienta KJ, Taichman RS (2013) Recruitment of mesenchymal stem cells into prostate tumours promotes metastasis. Nat Commun 4:1795

50. Kao CJ, Martiniez A, Shi XB, Yang J, Evans CP, Dobi A, Devere White RW, Kung HJ (2014) miR-30 as a tumor suppressor connects EGF/Src signal to ERG and EMT. Oncogene 33:2495–2503

51. Kelly SP, Anderson WF, Rosenberg PS, Cook MB (2017) Past, current, and future incidence rates and burden of metastatic prostate Cancer in the United States. Eur Urol Focus

52. Latulippe E, Satagopan J, Smith A, Scher H, Scardino P, Reuter V, Gerald WL (2002) Comprehensive gene expression analysis of prostate cancer reveals distinct transcriptional programs associated with metastatic disease. Cancer Res 62:4499–4506

53. Lichner Z, Ding Q, Samaan S, Saleh C, Nasser A, Al-Haddad S, Samuel JN, Fleshner NE, Stephan C, Jung K, Yousef GM (2015) miRNAs dysregulated in association with Gleason grade regulate extracellular matrix, cytoskeleton and androgen receptor pathways. J Pathol 237:226–237

54. Liu C, Kelnar K, Liu B, Chen X, Calhoun-Davis T, Li H, Patrawala L, YAN H, Jeter C, Honorio S, Wiggins JF, Bader AG, Fagin R, Brown D, Tang DG (2011) The microRNA miR-34a inhibits prostate cancer stem cells and metastasis by directly repressing CD44. Nat Med 17:211–215
55. Liu C, Liu R, Zhang D, Deng Q, Liu B, Chao HP, Rycaj K, Takata Y, Lin K, Lu Y, Zhong Y, Krolewski J, Shen J, Tang DG (2017) MicroRNA-141 suppresses prostate cancer stem cells and metastasis by targeting a cohort of pro-metastasis genes. Nat Commun 8:14270
56. Liu R, Liu C, Zhang D, Liu B, Chen X, Rycaj K, Jeter C, Calhoun-Davis T, Li Y, Yang T, Wang J, Tang DG (2016) miR-199a-3p targets stemness-related and mitogenic signaling pathways to suppress the expansion and tumorigenic capabilities of prostate cancer stem cells. Oncotarget 7:56628–56642
57. Liu W, Laitinen S, Khan S, Vihinen M, Kowalski J, Yu G, Chen L, Ewing CM, Eisenberger MA, Carducci MA, Nelson WG, Yegnasubramanian S, Luo J, Wang Y, Xu J, Isaacs WB, Visakorpi T, Bova GS (2009) Copy number analysis indicates monoclonal origin of lethal metastatic prostate cancer. Nat Med 15:559–565
58. Liu X, Chen Q, Yan J, Wang Y, Zhu C, Chen C, Zhao X, Xu M, Sun Q, Deng R, Zhang H, Qu Y, Huang J, Jiang B, Yu J (2013) MiRNA-296-3p-ICAM-1 axis promotes metastasis of prostate cancer by possible enhancing survival of natural killer cell-resistant circulating tumour cells. Cell Death Dis 4:e928
59. Liu YN, Yin J, Barrett B, Sheppard-Tillman H, Li D, Casey OM, Fang L, Hynes PG, Ameri AH, Kelly K (2015) Loss of androgen-regulated MicroRNA 1 activates SRC and promotes prostate Cancer bone metastasis. Mol Cell Biol 35:1940–1951
60. Msaouel P, Pissimissis N, Halapas A, Koutsilieris M (2008) Mechanisms of bone metastasis in prostate cancer: clinical implications. Best Pract Res Clin Endocrinol Metab 22:341–355
61. Nam RK, Benatar T, Wallis CJ, Amemiya Y, Yang W, Garbens A, Naeim M, Sherman C, Sugar L, Seth A (2016) MiR-301a regulates E-cadherin expression and is predictive of prostate cancer recurrence. Prostate 76:869–884
62. Nguyen DX, Bos PD, Massague J (2009) Metastasis: from dissemination to organ-specific colonization. Nat Rev Cancer 9:274–284
63. Patrawala L, Calhoun-Davis T, Schneider-Broussard R, Tang DG (2007) Hierarchical organization of prostate cancer cells in xenograft tumors: the CD44+alpha2beta1+ cell population is enriched in tumor-initiating cells. Cancer Res 67:6796–6805
64. Patrawala L, Calhoun T, Schneider-Broussard R, Li H, Bhatia B, Tang S, Reilly JG, Chandra D, Zhou J, Claypool K, Coghlan L, Tang DG (2006) Highly purified CD44+ prostate cancer cells from xenograft human tumors are enriched in tumorigenic and metastatic progenitor cells. Oncogene 25:1696–1708
65. Peng X, Guo W, Liu T, Wang X, Tu X, Xiong D, Chen S, Lai Y, Du H, Chen G, Liu G, Tang Y, Huang S, Zou X (2011) Identification of miRs-143 and -145 that is associated with bone metastasis of prostate cancer and involved in the regulation of EMT. PLoS One 6:e20341
66. Pereira DM, Rodrigues PM, Borralho PM, Rodrigues CM (2013) Delivering the promise of miRNA cancer therapeutics. Drug Discov Today 18:282–289
67. Pritchard CC, Mateo J, Walsh MF, de Sarkar N, Abida W, Beltran H, Garofalo A, Gulati R, Carreira S, Eeles R, Elemento O, Rubin MA, Robinson D, Lonigro R, Hussain M, Chinnaiyan A, Vinson J, Filipenko J, Garraway L, Taplin ME, Aldubayan S, Han GC, Beightol M, Morrissey C, Nghiem B, Cheng HH, Montgomery B, Walsh T, Casadei S, Berger M, Zhang L, Zehir A, Vijai J, Scher HI, Sawyers C, Schultz N, Kantoff PW, Solit D, Robson M, van Allen EM, Offit K, de Bono J, Nelson PS (2016) Inherited DNA-repair gene mutations in men with metastatic prostate Cancer. N Engl J Med 375:443–453
68. Qu Y, Li WC, Hellem MR, Rostad K, Popa M, Mccormack E, Oyan AM, Kalland KH, Ke XS (2013) MiR-182 and miR-203 induce mesenchymal to epithelial transition and self-sufficiency of growth signals via repressing SNAI2 in prostate cells. Int J Cancer 133:544–555

69. Ru P, Steele R, Newhall P, Phillips NJ, Toth K, Ray RB (2012) miRNA-29b suppresses prostate cancer metastasis by regulating epithelial-mesenchymal transition signaling. Mol Cancer Ther 11:1166–1173
70. Rycaj K, Li H, Zhou J, Chen X, Tang DG (2017) Cellular determinants and microenvironmental regulation of prostate cancer metastasis. Semin Cancer Biol 44:83–97
71. Sadeghi M, Ranjbar B, Ganjalikhany MR, Khan FM, Schmitz U, Wolkenhauer O, Gupta SK (2016) MicroRNA and transcription factor gene regulatory network analysis reveals key regulatory elements associated with prostate Cancer progression. PLoS One 11:e0168760
72. Saini S, Majid S, Shahryari V, Arora S, Yamamura S, Chang I, Zaman MS, Deng G, Tanaka Y, Dahiya R (2012) miRNA-708 control of CD44(+) prostate cancer-initiating cells. Cancer Res 72:3618–3630
73. Saini S, Majid S, Yamamura S, Tabatabai L, Suh SO, Shahryari V, Chen Y, Deng G, Tanaka Y, Dahiya R (2011) Regulatory role of mir-203 in prostate Cancer progression and metastasis. Clin Cancer Res 17:5287–5298
74. Selth LA, Das R, Townley SL, Coutinho I, Hanson AR, Centenera MM, Stylianou N, Sweeney K, Soekmadji C, Jovanovic L, Nelson CC, Zoubeidi A, Butler LM, Goodall GJ, Hollier BG, Gregory PA, Tilley WD (2017) A ZEB1-miR-375-YAP1 pathway regulates epithelial plasticity in prostate cancer. Oncogene 36:24–34
75. Shen PF, Chen XQ, Liao YC, Chen N, Zhou Q, Wei Q, Li X, Wang J, Zeng H (2014) MicroRNA-494-3p targets CXCR4 to suppress the proliferation, invasion, and migration of prostate cancer. Prostate 74:756–767
76. Siegel RL, Miller KD, Jemal A (2017) Cancer statistics, 2017. CA Cancer J Clin 67:7–30
77. Sottnik JL, Dai J, Zhang H, Campbell B, Keller ET (2015) Tumor-induced pressure in the bone microenvironment causes osteocytes to promote the growth of prostate cancer bone metastases. Cancer Res 75:2151–2158
78. Stankiewicz E, Mao X, Mangham DC, Xu L, Yeste-Velasco M, Fisher G, North B, Chaplin T, Young B, Wang Y, Kaur Bansal J, Kudahetti S, Spencer L, Foster CS, Moller H, Scardino P, Oliver RT, Shamash J, Cuzick J, Cooper CS, Berney DM, Lu YJ (2017) Identification of FBXL4 as a metastasis associated gene in prostate Cancer. Sci Rep 7:5124
79. Stewart DA, Cooper CR, Sikes RA (2004) Changes in extracellular matrix (ECM) and ECM-associated proteins in the metastatic progression of prostate cancer. Reprod Biol Endocrinol 2:2
80. Sung SY, Liao C H, Wu HP, Hsiao WC, Wu IH, Jinpu Yu, Lin SH, Hsieh CL (2013) Loss of let-7 microRNA upregulates IL-6 in bone marrow-derived mesenchymal stem cells triggering a reactive stromal response to prostate cancer. PLoS One 8:e71637
81. Tai HC, Chang AC, Yu HJ, Huang CY, Tsai YC, Lai YW, Sun HL, Tang CH, Wang SW (2014) Osteoblast-derived WNT-induced secreted protein 1 increases VCAM-1 expression and enhances prostate cancer metastasis by down-regulating miR-126. Oncotarget 5:7589–7598
82. Takeshita F, Patrawala L, Osaki M, Takahashi RU, Yamamoto Y, Kosaka N, Kawamata M, Kelnar K, Bader AG, Brown D, Ochiya T (2010) Systemic delivery of synthetic microRNA-16 inhibits the growth of metastatic prostate tumors via downregulation of multiple cell-cycle genes. Mol Ther 18:181–187
83. Tang DG, Patrawala L, Calhoun T, Bhatia B, Choy G, Schneider-Broussard R, JETER C (2007) Prostate cancer stem/progenitor cells: identification, characterization, and implications. Mol Carcinog 46:1–14
84. Tantivejkul K, Kalikin LM, Pienta KJ (2004) Dynamic process of prostate cancer metastasis to bone. J Cell Biochem 91:706–717
85. Tomlins SA, Rhodes DR, Perner S, Dhanasekaran SM, Mehra R, Sun XW, Varambally S, Cao X, Tchinda J, Kuefer R, Lee C, Montie JE, Shah RB, Pienta KJ, Rubin MA, Chinnaiyan AM (2005) Recurrent fusion of TMPRSS2 and ETS transcription factor genes in prostate cancer. Science 310:644–648
86. Tong AW, Fulgham P, Jay C, Chen P, Khalil I, Liu S, Senzer N, Eklund AC, Han J, Nemunaitis J (2009) MicroRNA profile analysis of human prostate cancers. Cancer Gene Ther 16:206–216

87. Tyekucheva S, Bowden M, Bango C, Giunchi F, Huang Y, Zhou C, Bondi A, Lis R, Van Hemelrijck M, Andren O, Andersson SO, Watson RW, Pennington S, Finn SP, Martin NE, Stampfer MJ, Parmigiani G, Penney KL, Fiorentino M, Mucci LA, Loda M (2017) Stromal and epithelial transcriptional map of initiation progression and metastatic potential of human prostate cancer. Nat Commun 8:420

88. Valencia K, Luis-Ravelo D, Bovy N, Anton I, Martinez-Canarias S, Zandueta C, Ormazabal C, Struman I, Tabruyn S, Rebmann V, de Las Rivas J, Guruceaga E, Bandres E, Lecanda F (2014) miRNA cargo within exosome-like vesicle transfer influences metastatic bone colonization. Mol Oncol 8:689–703

89. Varambally S, Dhanasekaran SM, Zhou M, Barrette TR, Kumar-Sinha C, Sanda MG, Ghosh D, Pienta KJ, Sewalt RG, Otte AP, Rubin MA, Chinnaiyan AM (2002) The polycomb group protein EZH2 is involved in progression of prostate cancer. Nature 419:624–629

90. Watahiki A, Wang Y, Morris J, Dennis K, O'dwyer HM, Gleave M, Gout PW, Wang Y (2011) MicroRNAs associated with metastatic prostate cancer. PLoS One 6:e24950

91. Xie H, Li L, Zhu G, Dang Q, Ma Z, He D, Chang L, Song W, Chang HC, Krolewski JJ, Nastiuk KL, Yeh S, Chang C (2015) Infiltrated pre-adipocytes increase prostate cancer metastasis via modulation of the miR-301a/androgen receptor (AR)/TGF-beta1/Smad/MMP9 signals. Oncotarget 6:12326–12339

92. Xue M, Liu H, Zhang L, Chang H, Liu Y, Du S, Yang Y, Wang P (2017) Computational identification of mutually exclusive transcriptional drivers dysregulating metastatic microRNAs in prostate cancer. Nat Commun 8:14917

93. Zoni E, van der Horst G, van de Merbel AF, Chen L, Rane JK, Pelger RC, Collins AT, Visakorpi T, Snaar-Jagalska BE, Maitland NJ, van der Pluijm G (2015) miR-25 modulates invasiveness and dissemination of human prostate Cancer cells via regulation of alphav- and alpha6-integrin expression. Cancer Res 75:2326–2336

Chapter 6
Epithelial-Mesenchymal Transition (EMT) and Prostate Cancer

Valerie Odero-Marah, Ohuod Hawsawi, Veronica Henderson, and Janae Sweeney

Abstract Typically the normal epithelial cells are a single layer, held tightly by adherent proteins that prevent the mobilization of the cells from the monolayer sheet. During prostate cancer progression, the epithelial cells can undergo epithelial-mesenchymal transition or EMT, characterized by morphological changes in their phenotype from cuboidal to spindle-shaped. This is associated with biochemical changes in which epithelial cell markers such as E-cadherin and occludins are down-regulated, which leads to loss of cell-cell adhesion, while mesenchymal markers such as vimentin and N-cadherin are up-regulated, thereby allowing the cells to migrate or metastasize to different organs. The EMT transition can be regulated directly and indirectly by multiple molecular mechanisms including growth factors and cytokines such as transforming growth factor-beta (TGF-β), epidermal growth factor (EGF) and insulin-like growth factor (IGF), and signaling pathways such as mitogen-activated protein kinase (MAPK) and Phosphatidylinositol 3-Kinase (PI3K). This signaling subsequently induces expression of various transcription factors like Snail, Twist, Zeb1/2, that are also known as master regulators of EMT. Various markers associated with EMT have been reported in prostate cancer patient tissue as well as a possible association with health disparities. There has been consideration to therapeutically target EMT in prostate cancer patients by targeting the EMT signaling pathways.

Keywords Epithelial-Mesenchymal Transition · Prostate Cancer · Transcription Factors · Growth Factors · Cytokines

V. Odero-Marah (✉) · O. Hawsawi · V. Henderson · J. Sweeney
Department of Biology, Clark Atlanta University, Atlanta, GA, USA
e-mail: voderomarah@cau.edu

© Springer International Publishing AG, part of Springer Nature 2018
H. Schatten (ed.), *Cell & Molecular Biology of Prostate Cancer*,
Advances in Experimental Medicine and Biology 1095,
https://doi.org/10.1007/978-3-319-95693-0_6

6.1 Introduction

In most metazoans, prostate epithelial cells are in close contact to the basal membrane, held together by tight junction and adherens junction proteins. However, during development, cuboidal epithelial cells undergo morphological and biochemical changes to transition into a mesenchymal phenotype which are more elongated and spindle-shaped. This process is called epithelial-mesenchymal transition or EMT, and can be divided into three different types primary, secondary and tertiary. The primary EMT takes place during early development and is well recognized at early gastrulation and neural crest development [1]. Gastrulation is described as the early formation of the three germ line layers (ectoderm, mesoderm, and endoderm) from the initial epithelial cells [2]. The post gastrulation is considered as the secondary EMT type, leading to formation of neural crest within ectodermal zones, thus giving rise to different cells such as neurons, bone, and mesodermal cells. At this point, the cells convert into epithelial type again by the reverse process of EMT called mesenchymal-epithelial transition (MET) [3]. The tertiary type of EMT can be well explained through a successive cycle of heart formation. During cardiac development, the mesodermal cells differentiate with other cardiac progenitors into two epithelial layers; another EMT process follows to form endothelial cell linings of the heart. The endothelial cells from atrioventricular canal undergo a tertiary EMT to form the endocardial cushion and later, the cells will assemble to form atrioventricular valvuloseptal complex [4].

Mesenchymal cells exhibit a front back end polarity with loss of structured cuboidal shape, and acquisition of mesenchymal markers which make this type of cells migratory, invasive and more resistant to apoptosis [5]. Molecularly, EMT is associated with loss of epithelial markers such as E-cadherin, occludin and zonula-occludens (ZO-1), and acquisition of mesenchymal markers such as vimentin, N-cadherin, and fibronectin [6].

Most patients with prostate cancer succumb to the disease due to the primary tumor metastasizing to an organ critical for survival such as the lungs or the liver [7]. Prostate cancer also has a propensity to metastasize to the bone [7]. Cancer cells have hijacked the EMT process to become invasive, migratory and acquire the ability to breakdown the basement membrane and metastasize (Fig. 6.1). However, not all the tumor cells are able to escape the primary organ and this phenomenon appears only in a specific population of the tumor cells [1]. EMT plays a critical role in cancer progression and metastasis [8]. Although the complete evidence of how the cancer cells undergo EMT is still ambiguous, strong evidence shows this process can be reproduced in animal models, including animal models of prostate cancer [9]. EMT is not characterized by a complete change in the cell identity, but more by a transient change in the cells' mobility and behavior. In tumors, incomplete EMT occurs where the cancer cells gain the mesenchymal characters while still expressing some epithelial markers, thus, without facing the complete transition as found within the embryo [9]. The majority of the death cases with prostate cancer are due

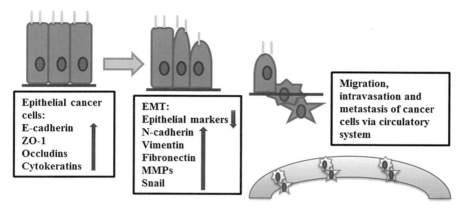

Fig. 6.1 Epithelial-mesenchymal transition (EMT) in cancer cells. Cuboidal epithelial cells can transition into spindle-shaped mesenchymal cells which is associated with downregulation of epithelial markers such as E-cadherin, zonula occludens-1 (ZO-1), occludins and cytokeratins, and metalloproteinases (MMPs) and Snail

to metastatic disease that does not respond to treatment, and that have become castration-resistant [10]. Androgenic /androgen receptor (AR) signaling plays a role not only in prostate organ development in early stages, but studies show that in the initial stages of tumorigenesis cancer cells depend on androgen to promote cell growth and inhibit apoptosis, but with androgen-deprivation therapy, some tumors with time become resistant and eventually metastatic [11]. EMT plays a critical role in the development of the metastatic castration resistant prostate cancer (mCRPC) [12]. Additionally, it has been reported that AR can repress E-cadherin and induce EMT; Liu et al., demonstrated that active AR is able to downregulate E-cadherin expression which led to loss of cell-cell adhesion and promotion of metastasis [13].

6.2   Transcription Factors that Regulate EMT

EMT can be induced by various transcription factors such as Snail, Slug and Twist [14]. Deficiency of Snail in the embryo leads to unsuccessful completion of the EMT process [15]. Snail transcription factor is a zinc finger protein, known as a master protein which regulates EMT. Snail regulates EMT by downregulating E-cadherin during both development and tumor progression [16]. Snail can regulate E-cadherin by binding to the E-box region within the E-cadherin promoter and repressing transcription in prostate cancer cells [17]. In prostate tumorigenesis, the high expression of Snail is associated with loss of E-cadherin [18]. In addition, Snail can also repress epithelial markers such as occludin and ZO-1 [17].

6.3   Growth Factors and Cytokines that Induce EMT in Prostate Cancer

Various growth factors and cytokines have been shown to contribute to the process of EMT in prostate cancer. Some of the growth factors and cytokines reported to play a role during prostate cancer progression are transforming growth factor beta (TGF-β), Insulin-like growth factors (IGF), epidermal growth factor (EGF), and CX3CL-1 [19–25]. The mechanism of action is through activation of growth factors and cytokines to their respective receptors leading to induction of signaling pathways downstream [26, 27].

Growth Factors and cytokines are secreted glycoproteins that act as signaling molecules to regulate various cellular functions [28]. The two words are often used interchangeably however, growth factors are assumed to have a positive role on cell proliferation whereas as cytokines can also have a negative effect on cell growth [28]. Some of the growth factors and downstream signaling pathways that regulate EMT in prostate cancer are shown in Fig. 6.2. One of the well-studied cytokines that plays a key role during tumor progression and metastasis is TGF-β. It has three family members namely, TGF-β1, TGF-β2, and TGF-β3 [29]. TGF-β has opposing roles during prostate cancer progression, as a tumor suppressor during the early disease stages and a tumor promoter in the later stages [21]. In the benign stages of prostate cancer, TGF-β binds to its receptors and activates its signaling pathway that leads to apoptosis [21]. It also mediates processes such as cell differentiation, cell proliferation

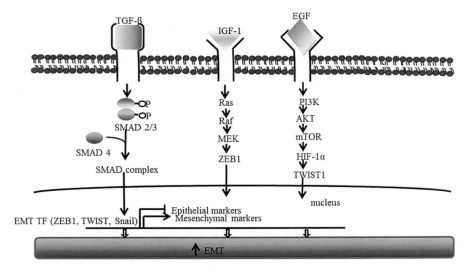

Fig. 6.2 Growth factor signaling pathways that triggers EMT in prostate cancer. Growth factors such as TGF-β, IGF-1 and EGF can trigger downstream signaling pathways such as MAPK and PI3K, that lead to activation of transcription factors (TF) such as Snail, ZEB1, TWIST. This eventually leads to downregulation of epithelial markers and upregulation of mesenchymal markers

and migration [21]. In late-stages of prostate cancer, TGF-β is shown to be up-regulated leading to increased cell invasion and metastasis [30]. It plays a role during EMT by downregulating epithelial markers such as E-cadherin and upregulating mesenchymal markers such as vimentin [31]. TGF-β cell signaling utilizes either a SMAD or non-SMAD pathway [21]. For the SMAD mediated pathway, TGF-β binds to its receptor, TGF-β type II receptor (TβRII) which leads to recruitment and activation of TβRI by phosphorylation at the serine and threonine residues [21]. The activated TβRI then recruits and phosphorylates SMAD 2 and SMAD 3 [21]. These two proteins then form complexes with SMAD4 leading to translocation into the nucleus where they regulate their target genes [21]. Examples of TGF-β signaling target genes are SMAD 7, p21, c-Jun, among others [19]. Thahur et al., showed that c-Jun binds to Snail promoter hence initiating migration and invasion of prostate cancer cells [19]. Some of the non-SMAD pathways are MAPK, mTOR, Ras, c-Src, PI3K/ AKT, RhoA, Cofilin, among others [21].

In most late stage tumors, TGF-β signaling components are lost or there are alterations in a downstream signaling component such as Ras activation [30]. One mechanism by which TGF-β signaling is altered in prostate cancer is through loss of TβRII, and this has been correlated with high grade tumors. Tu et al. did a study using transgenic mice with a TβRII mutation (DNIIR) that rendered it a dominant negative mutant [20]. They observed that the mutant mice had increased tumor metastasis compared to control mice [20], thus demonstrating that loss of TGF-β signaling is one mechanism by which it acquires its tumor promoter role in late stage prostate cancer [20]. Therapeutic treatments designed to target TGF-β signaling should seek to keep its apoptotic role while inhibiting the tumor invasion and metastasis role [30].

Insulin-like Growth Factor (IGF) is a growth factor that is known to regulate differentiation, apoptosis, proliferation, and cellular metabolism [22]. It has been implicated in prostate cancer bone metastasis [22]. IGF has two family members, IGF-I and IGF-II and two receptors, IGF-IR and IGF-IIR, as well as 6 binding proteins (IGFBPs 1–6) [22]. These proteins interact with each other as well as crosstalk with other signaling pathways [22]. IGF-IR is a tyrosine kinase receptor located on the cell membrane [22]. When IGF binds to its receptor it induces downstream signaling pathways such as mitogen-activated protein kinase (MAPK) and Phosphatidylinositol 3- Kinase (PI3K) [22]. Insulin-like Growth Factor (IGF) is a growth factor that has been reported to increase EMT in prostate cancer [23]. This occurs by up-regulation of ZEB1 expression which is a transcription factor known to down-regulate E-cadherin levels [23]. In this study, they treated ARCAPE prostate cancer cells with recombinant IGF-1 and showed that ZEB1 was increased two-fold in the nucleus compared to the control, leading to increased MAPK activation and cell migration [23].

Another growth factor that plays a role in the process of EMT in prostate cancer is Epidermal Growth Factor (EGF) [32]. Lorenzo et al., did a clinical study of prostate cancer patients and assessed Epidermal Growth Factor Receptor (EGFR) expression. Their results showed that EGFR was expressed in all the patients they assessed who had metastasis [24]. EGF has been reported to induce EMT through increased expression of transcription factors responsible for reducing E-cadherin and promoting cancer invasion [32]. They showed a mechanism in which EGF increases prostate cancer progression through a Ras/ STAT3/ HIF-1 alpha/ TWIST1/ N-cadherin signaling pathway [32].

6.4   Clinical Evidence of EMT in Prostate Cancer

Epithelial mesenchymal transition (EMT) is a distinguishable feature of aggressive tumors in prostate cancer. In prostate cancer, several transcription factors are instrumental in inducing EMT such as *Snail* and *Twist*. Following regulation by *Snail*, EMT occurs and prostate cancer cells experience reduced E-cadherin and increased/ up regulation of N-cadherin [33]. In the 2007 study aimed to determine the significance of EMT, tissue from a consecutive series of 104 men treated by radical prostatectomy for clinically localized cancer during 1988–1994 was utilized [33]. The tissue microarray was studied using immunohistochemistry techniques to analyze cell adhesion molecules including classic cadherins (E-cadherin, N-cadherin, and P-cadherin) and β-Catenin and p120^{CTN} and confirmed using Western blot analysis. In this study, it was determined that the decrease of E-cadherin and subsequent up regulation of N-cadherin (E-cadherin to N-Cadherin switch suggestive of EMT) is a strong predictor of clinical recurrence after radical prostatectomy [33]. This finding is a direct indicator that cell adhesion molecules may be used as prognostic information along with histologic evaluation and also demonstrates the importance of EMT for patient prognosis of human prostate cancer [33].

In other clinically related research, tissue (archived, formalin fixed, and paraffin-embedded) containing both tumor and adjacent normal tissue was obtained from surgically resected prostate cancer specimens (10 primary and 10 prostate cancer bone metastasis). Each tissue section was immunostained using specific antibodies for EMT biomarkers E-cadherin, Nuclear factor kappa B (NF-κB), Notch-1, ZEB1, and Platelet-derived growth factor D (PDGF-D) [34]. Slides of each marker were scored and accessed by stain localization, intensity, and percentage of stained cells within the tumors. From the 20 samples of primary and bone metastasis, E-cadherin was expressed within the membrane; Vimentin and PDGF-D expression in the cytoplasm; and NF-κB, Notch-1, and ZEB1 were expressed in the nucleus [34]. Results in this study demonstrated that the upregulation of all observed EMT markers, specifically Notch-1 play a significant role in prostate cancer and bone metastasis [34].

6.5   EMT in Prostate Cancer Health Disparities

Among men, prostate cancer is the most diagnosed cancer as well as the second leading cause of death [35]. African American men have a two-fold increase in mortality due to prostate cancer as compared to Caucasian men [35]. Some have suggested that this health disparity could be due to biological factors. To date it has been difficult to find data on EMT in prostate cancer health disparities. However, research has been conducted on Kaiso, a transcriptional factor that is a member of the BTB/POZ zinc finger protein family and can induce EMT. Localization of Kaiso in the cell is characterized by a methylation-dependent silencing of E-cadherin [36] . As with regulation of EMT by *Snail* [33], down regulation of E-cadherin by Kaiso is associated with increased cell migration invasiveness and tumor aggressiveness [36]. Specifically, it has been observed that a shift in localization from the cytoplasm to the nucleus in cells causes methylation-dependent silencing of E-cadherin, which promotes cell migration and aggressiveness [36]. Experimentation was conducted to determine the relationship between Kaiso and miR-31 in a panel of cells: normal cell line (PREC), immortal normal epithelial cell line (RC-77 N/E), and Caucasian human prostate cancer lines LNCaP, DU-145, C4-2B and PC-3 [36]. MiR-31 is a microRNA that plays a role in cell proliferation, and EMT. Quantitative real-time polymerase chain reaction (qRT-PCR) revealed that Kaiso expression was low in PREC and RC-77 N/E but higher in prostate cancer cell lines with expression increasing in more aggressive cells like PC-3 and C4-2B cells. In the panel of prostate cancer cells, Kaiso levels were negatively/inversely correlated with miR-31 expression [36]. These results were supported in the observation that patients with high Kaiso levels and low miR-31 expression experienced the most significant decrease in survival compared to patients who exhibited low mRNA Kaiso levels with high miR-31 expression, and that the expression of Kaiso was higher in African American patient tissue as compared to Caucasian American patient tissue [36]. More studies are needed in the area of EMT in prostate cancer health disparities.

6.6 Therapeutic Targeting of EMT in Prostate Cancer

Biomarkers including Snail, E-cadherin, N-cadherin, Vimentin, ZEB1, TWIST have been demonstrated to play a role in the upregulation of EMT. Other EMT regulatory factors include castration, and androgen deprivation [37]. An N-cadherin antagonist, Alcohol dehydrogenase-1 (ADH-1), is a targeted therapy for EMT that has been proposed [38]. However, larger studies must be done to validate their findings. Therapeutic strategies that intervene the EMT process or reverse EMT phenotypes may be alternatives for cancer therapy. Specific N-cadherin antibodies can suppress the up regulation of EMT simultaneously decreasing tumor growth invasion and migration and blocking the progression to castration-resistance [37].

Small molecule inhibitors are also being tested as possible therapies that target EMT. For example, one compound, DZ-50, was shown to inhibit EMT in prostate cancer cells by targeting the TGF-β and IGF axis [39]. Another potent small-molecule compound, BMS-345541, was identified as a highly selective IKKα and IKKβ inhibitor that could inhibit EMT in prostate cancer cells and induce apoptosis [40].

Targeting the growth factors that promote EMT have also been studied preclinically and in clinical trials. However, results for IGF-1R inhibitors as single agents in prostate cancer clinical trials have not been promising [41]. Neutralizing antibodies, antisense oligonucleotides and small molecule inhibitors have also been tested in pre-clinical studies to target tumor-promoting activities of TGF-β [42].

Natural products have also been proposed as potential therapies for prostate cancer EMT. Studies have shown that muscadine grape skin extract (MSKE) that has strong anti-oxidant activity, can inhibit EMT in prostate cancer cells and promote apoptosis without affecting normal cells [43, 44]. This product is also being tested in clinical trials in prostate cancer patients [45].

6.7   Conclusions

Prostate cancer cells have hijacked the EMT process to become invasive, migratory and metastatic. This EMT can be induced by various growth factors, cytokines and downstream signaling leading to activation of various transcription factors. Evidence of EMT has also been shown in prostate cancer patients. Therefore, some of these growth factor- and cytokine-mediated pathways provide excellent targets for therapeutic interventions for treatment of prostate cancer patients *via* antagonizing EMT.

References

1. Yang J, Weinberg RA (2008) Epithelial-mesenchymal transition: at the crossroads of development and tumor metastasis. Dev Cell 14(6):818–829
2. Thiery JP et al (2009) Epithelial-mesenchymal transitions in development and disease. Cell 139(5):871–890
3. Kalluri R, Weinberg RA (2009) The basics of epithelial-mesenchymal transition. J Clin Invest 119(6):1420
4. Imran Khan M et al (2015) Role of epithelial mesenchymal transition in prostate tumorigenesis. Curr Pharm Des 21(10):1240–1248
5. Thiery JP (2002) Epithelial–mesenchymal transitions in tumour progression. Nat Rev Cancer 2(6):442–454
6. Micalizzi DS, Farabaugh SM, Ford HL (2010) Epithelial-mesenchymal transition in cancer: parallels between normal development and tumor progression. J Mammary Gland Biol Neoplasia 15(2):117–134
7. Nauseef JT, Henry MD (2011) Epithelial-to-mesenchymal transition in prostate cancer: paradigm or puzzle? Nat Rev Urol 8(8):428–439
8. Radisky DC (2005) Epithelial-mesenchymal transition. J Cell Sci 118(19):4325–4326
9. Nieto MA, Cano A (2012) The epithelial–mesenchymal transition under control: global programs to regulate epithelial plasticity. In Seminars in cancer biology. Elsevier
10. Sethi S et al (2011) Molecular signature of epithelial-mesenchymal transition (EMT) in human prostate cancer bone metastasis. Am J Transl Res 3(1):90
11. Harris WP et al (2009) Androgen deprivation therapy: progress in understanding mechanisms of resistance and optimizing androgen depletion. Nat Clin Pract Urol 6(2):76–85

12. Grant CM, Kyprianou N (2013) Epithelial mesenchymal transition (EMT) in prostate growth and tumor progression. Translational andrology and urology 2(3):202
13. Liu Y-N et al (2008) Activated androgen receptor downregulates E-cadherin gene expression and promotes tumor metastasis. Mol Cell Biol 28(23):7096–7108
14. Cano A et al (2000) The transcription factor snail controls epithelial–mesenchymal transitions by repressing E-cadherin expression. Nat Cell Biol 2(2):76
15. Nieto MA (2002) The snail superfamily of zinc-finger transcription factors. Nat Rev Mol Cell Biol 3(3):155–166
16. Peinado H, Olmeda D, Cano A (2007) Snail, Zeb and bHLH factors in tumour progression: an alliance against the epithelial phenotype? Nat Rev Cancer 7(6):415–428
17. Smith BN, Odero-Marah VA (2012) The role of snail in prostate cancer. Cell Adhes Migr 6(5):433–441
18. Zhau HE et al (2008) Epithelial to mesenchymal transition (EMT) in human prostate cancer: lessons learned from ARCaP model. Clin Exp Metastasis 25(6):601
19. Thakur N, Gudey S, Marcusson A, Fu J, Bergh A, Heldin C, Landstrom M (2014) TGF-ß-induced invasion of prostate Cancer cells is promoted by c-Jun-dependent transcriptional activation of Snail1. Cell Cycle 13(15):2400–2414
20. Tu WH et al (2003) The loss of TGF-beta signaling promotes prostate cancer metastasis. Neoplasia 5(3):267–277
21. Cao Z, Kyprianou N (2015) Mechanisms navigating the TGF-ß pathway in prostate cancer. Asian J Urol 2:11–18
22. Gennigens C, Menetrier-Caux C, Droz JP (2006) Insulin-like growth factor (IGF) family and prostate cancer. Crit Rev Oncol Hematol 58(2):124–145
23. Graham TR et al (2008) Insulin-like growth factor-I-dependent up-regulation of ZEB1 drives epithelial-to-mesenchymal transition in human prostate cancer cells. Cancer Res 68(7):2479–2488
24. Di Lorenzo G et al (2002) Expression of epidermal growth factor receptor correlates with disease relapse and progression to androgen-independence in human prostate cancer. Clin Cancer Res 8(11):3438–3444
25. Tang J et al (2016) CX3CL1 increases invasiveness and metastasis by promoting epithelial-to-mesenchymal transition through the TACE/TGF-alpha/EGFR pathway in hypoxic androgen-independent prostate cancer cells. Oncol Rep 35(2):1153–1162
26. Grotzinger J (2002) Molecular mechanisms of cytokine receptor activation. Biochim Biophys Acta 1592(3):215–223
27. Maruyama IN (2014) Mechanisms of activation of receptor tyrosine kinases: monomers or dimers. Cell 3(2):304–330
28. Hughes FJ et al (2006) Effects of growth factors and cytokines on osteoblast differentiation. Periodontol 2000 41:48–72
29. Poniatowski LA et al (2015) Transforming growth factor Beta family: insight into the role of growth factors in regulation of fracture healing biology and potential clinical applications. Mediat Inflamm 2015:137823
30. Akhurst RJ, Derynck R (2001) TGF-beta signaling in cancer--a double-edged sword. Trends Cell Biol 11(11):S44–S51
31. Xu J, Lamouille S, Derynck R (2009) TGF-beta-induced epithelial to mesenchymal transition. Cell Res 19(2):156–172
32. Cho KH et al (2014) A ROS/STAT3/HIF-1alpha signaling cascade mediates EGF-induced TWIST1 expression and prostate cancer cell invasion. Prostate 74(5):528–536
33. Gravdal K et al (2007) A switch from E-cadherin to N-cadherin expression indicates epithelial to mesenchymal transition and is of strong and independent importance for the progress of prostate cancer. Clin Cancer Res 13(23):7003–7011
34. Sethi S et al (2010) Molecular signature of epithelial-mesenchymal transition (EMT) in human prostate cancer bone metastasis. Am J Transl Res 3(1):90–99
35. Miller DB (2014) Pre-screening age African-American males: what do they know about prostate cancer screening, knowledge, and risk perceptions? Soc Work Health Care 53(3):268–288

36. Wang H et al (2016) Kaiso, a transcriptional repressor, promotes cell migration and invasion of prostate cancer cells through regulation of miR-31 expression. Oncotarget 7(5):5677–5689

37. Li P, Yang R, Gao W-Q (2014) Contributions of epithelial-mesenchymal transition and cancer stem cells to the development of castration resistance of prostate cancer. Mol Cancer 13(1):55

38. Makrilia N et al (2009) Cell adhesion molecules: role and clinical significance in cancer. Cancer Investig 27(10):1023–1037

39. Cao Z et al (2017) Reversion of epithelial-mesenchymal transition by a novel agent DZ-50 via IGF binding protein-3 in prostate cancer cells. Oncotarget 8(45):78507–78519

40. Ping H et al (2016) IKK inhibitor suppresses epithelial-mesenchymal transition and induces cell death in prostate cancer. Oncol Rep 36(3):1658–1664

41. Qu X et al (2017) Update of IGF-1 receptor inhibitor (ganitumab, dalotuzumab, cixutumumab, teprotumumab and figitumumab) effects on cancer therapy. Oncotarget 8(17):29501–29518

42. Jones E, Pu H, Kyprianou N (2009) Targeting TGF-beta in prostate cancer: therapeutic possibilities during tumor progression. Expert Opin Ther Targets 13(2):227–234

43. Burton LJ et al (2014) Muscadine grape skin extract reverts snail-mediated epithelial mesenchymal transition via superoxide species in human prostate cancer cells. BMC Complement Altern Med 14:97

44. Hudson TS et al (2007) Inhibition of prostate cancer growth by muscadine grape skin extract and resveratrol through distinct mechanisms. Cancer Res 67(17):8396–8405

45. Paller CJ et al (2015) A phase I study of muscadine grape skin extract in men with biochemically recurrent prostate cancer: safety, tolerability, and dose determination. Prostate 75(14):1518–1525

Chapter 7
The Role of Multi-Parametric MRI and Fusion Biopsy for the Diagnosis of Prostate Cancer – A Systematic Review of Current Literature

Debashis Sarkar

Abstract

Introduction

The use of mutiparametric MRI (MpMRI) guided fusion biopsy is becoming an increasingly popular investigation in an aid to increase diagnostic yield in those suspected of having prostate cancer (PCa). Before adopting this technology, it is necessary to confirm the accuracy, so that PCa can be reliably diagnosed with characterisation.

Materials and Methods

This chapter analysed the evidences, which varied from well-designed randomised controlled trials to case series to detect the accuracy of MpMRI compared with biopsy/ histology.

Results

MpMRI incorporating T2 and diffusion weighted imaging only detects tumours in around 92% cases. When dynamic contrast enhancement is added, cancer diagnosis is significantly improved. Fusion biopsy increases the detection of high-risk PCa by 32% over conventional biopsy alone.

Conclusion

This review also revealed that fusion biopsy did not increase cancer detection rate but combined biopsy (Systematic and fusion) provide the highest detection rate for the diagnosis of PCa.

Keywords Multi-parametric magnetic resonance imaging · Prostate biopsy Prostate cancer · Histology · Fusion biopsy

D. Sarkar (✉)
ST4, Urology Department, St Richard Hospital, Chichester, UK

© Springer International Publishing AG, part of Springer Nature 2018   111
H. Schatten (ed.), *Cell & Molecular Biology of Prostate Cancer*,
Advances in Experimental Medicine and Biology 1095,
https://doi.org/10.1007/978-3-319-95693-0_7

Abbreviations

DCE MRI	Dynamic Contrast Enhanced MRI
DW MRI	Diffusion weighted MRI
MpMRI	Multi-parametric MRI
TB	Targeted biopsy/Fusion biopsy
TPSB	Transperineal Saturation biopsy
TRUS Biopsy	Transrectal Ultrasound guided biopsy

7.1 Introduction

Prostate cancer is a one of the most common cancers in men in the western world. It was the second most common cause of cancer death in men (Prostate cancer research, UK) in the UK. Over the last 35 years prostate cancer incidence rates in Great Britain have more than tripled, however much of this is attributed to increased detection with widespread use of serum Prostate Specific Antigen (PSA) testing (1). In Europe, around 417,000 new cases of prostate cancer were estimated to have been diagnosed in 2012 [1].

The patient's history, physical examination including digital rectal examination (DRE) and serum PSA are the triggering factors for transrectal ultrasound guided prostate biopsy (TRUS)/ systematic biopsy. DRE is a crude tool with variability from clinician to clinician and has a low predictive value [2]. Sensitivity and specificity of PSA is controversial. The conventional TRUS guided (10–12 core) systematic biopsy also fails to detect PCa in up to 25% of cases [3]. Therefore, suspicion of malignancy remains in a significant number of men, especially if PSA is persistently raised or DRE is abnormal or in initial biopsy with a typical acinar small cell proliferation (ASAP)/ high-grade PIN. Some patients undergo numerous repeat negative conventional biopsies over several years, subjecting them to anxiety and discomfort, with an associated added cost. The optimum management of this group is unclear. Transperineal/transrectal saturation (>20 core) biopsy of prostate has been reported to detect and map out cancer in 23–47% of men requiring repeat biopsy, but with a complication rate of urinary retention in 11–39% [4, 5].

In recent years, use of multi-parametric MRI (MpMRI) and fusion prostate biopsy has become an increasingly popular choice of investigation, as few targeted cores are needed to confirm the diagnosis [Figs. 7.1–7.2]. MpMRI has been used since 2005 to better identify and characterise PCa [6]. The functional sequences of MpMRI parameters are T2 weighted image, dynamic contrast-enhancement (DCE) and diffusion-weighted imaging (DWI), including the calculation of apparent diffusion co-efficient (ADC) maps. Another parametre, MR spectroscopy has recently fallen out of favour. If more than one parameter is used in MRI then it is called Multi-parametric MRI and PI-RADS scoring system was adopted to characterize the lesions on MpMRI. (Score 1-Extremely unlikely, 2- Unlikely, 3-Equivocal, 4-Likely, 5-Extremely likely for the lesion). MpMRI guided fusion biopsy can

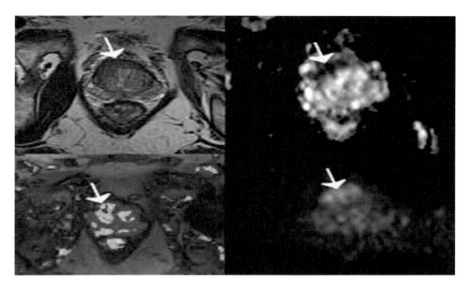

Fig. 7.1 (Taken from-Pedler K.et al. 2015). Multi-parametric images of an anterior prostate tumour. Left upper: T2 weighted; right upper: diffusion weighted coefficient map; left lower corner: dynamic contrast enhanced map; right lower corner: diffusion-weighted image map

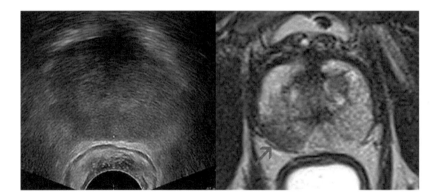

Fig. 7.2 (Taken from Fusion Guided Biopsy: A smarter way to look for prostate cancer) The MRI and ultrasound-fused image

detect more significant prostate cancers that are missed by conventional biopsy [7, 8]. MRI-USS fusion biopsy uses software that fuses stored MRI with real-time ultrasound (MRI-US) [Fig. 7.3]. The correlation between biopsy and final prostate pathology has been improved by MRI-guided biopsy as compared to TRUS guided prostate biopsy alone [9].

For these reasons, multi-parametric MRI and fusion biopsy are marketed as an emerging tool in prostate cancer diagnosis, as many patients don't wish to undergo a repeat conventional biopsy or saturation biopsy to confirm a possible diagnosis.

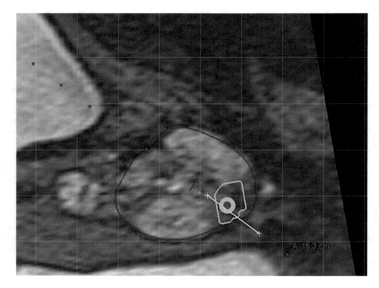

Fig. 7.3 (Taken from-Fusion Guided Biopsy: A smarter way to look for prostate cancer)- Fusion guided image, have seen during the biopsy procedure, with the prostate outlined in red, the suspected tumor in green and the biopsy needle in yellow

7.2   Aims of this Chapter

In patients with a negative conventional TRUSB but on-going suspicion of prostate cancer, the next line of investigation requires for definitive diagnosis or exclusion of malignancy, in order to prevent further uncertainty.

A negative MpMRI has been proposed as reasonable exclusion criteria for performing a repeat TRUSB/TPSB in many studies and a positive MpMRI can act as a trigger for repeat biopsy and in this way many repeat biopsies can be prevented. The patient with a positive lesion on MRI can undergo MRI-USS fusion biopsy to increase the diagnostic yield.

To examine the accuracy of multi-parametric MRI comparing with standard systematic prostate biopsy (10–12 cores), fusion biopsy (1–4 core), and final prostate pathology for either initial diagnosis or in those who have had one or more sets of negative conventional prostate biopsy, but in whom PCa is still suspected, a systematic approach was performed to reach a final conclusion in this chapter.

7.3   Systematic Literature Search Strategy

7.3.1   Introduction

The aim of this literature search is to obtain as many relevant current citations as possible in order to make a reasoned and unbiased judgment regarding the accuracy of multi-parametric MRI.

7.3.2   Search Methodology

The first part of this methodology is the formulation of this chapter's question in detail, which will aid the formulation of the search strategy undertaken to facilitate the retrieval of most current evidence. This will then be followed with use of diagrams of various keywords and combinations of keywords, derived from this question, to commence the literature search. Electronic databases to obtain current and relevant evidence, which is detailed later in this chapter, were used in a structured manner to enable reproducibility of the literature search.

In order to facilitate the literature search, the mnemonic PICO format (Table 7.1, 7.2) was used to help formulate a question, which in turn would aid developing a search strategy and therefore retrieval of relevant clinical evidence [10]. A "well - built" question consists of four parts- patient problem/ population, intervention, comparison and outcome. By expanding each component, appropriate

Table 7.1   Acronym PICO to formulate a clinical question

Patient problem / population of interest	Repeat prostate biopsy due to ongoing high PSA or abnormal DRE or negative conventional biopsy.
Intervention	Multi-parametric MRI
Comparison of interest	Compare with prostate Biopsy/histology
Outcome of interest	To prevent many repeat biopsies

Table 7.2   Search strategy

	Searches	Results
1	Prostate cancer and biopsy	8483
2	Prostate cancer and multi parametric MRI	15
3	1 or 2	8488
4	Control and trial	172,748
5	Randomised and controlled and trial	43,173
6	4 or 5	204,570
7	3 and 6	224
8	Limit 7 to (English language, humans, year = "2010-current")	103

search terms will be determined which would help develop an efficient approach to the question.

The search interval was 2010–2016, limited to articles published in English and on humans. A literature search was then conducted using Ovid Medline, PubMed and the Cochrane Database for Systematic Reviews for the key words used "Multi-parametric MRI" AND "Prostate Cancer" AND "Efficacy of MRI" AND" Prostate Biopsy". Only papers investigating the efficacy of MpMRI were included. The literature search revealed a large number of studies including randomised controlled trials (RCT), case series and review articles. Ideally meta-analysis and systematic reviews would have been ideal as these are all high levels of evidence, however only two suitable were identified. This may be due to the relatively new development of MpMRI resulting in their being an inadequate number of RCT's or the RCT may not be necessary for investigating the efficacy of this investigation.

Initially it seemed ideal to consider UK and non-UK based studies, as this would allow a global comparison of attitudes and trends in prostatic cancer investigation and diagnostic methods. A total of 16 studies (2 RCT's and 1 systematic review and 1 meta analysis included) were selected for this review to answer this question in this chapter.

7.4 Results– (Detailed Result of the Studies In– Table 7.3)

These studies concluded that MpMRI with T2W images and Diffusion-Weighted images (DWI) could detect PCa in 92% cases and when Dynamic Contrast Enhancement (DCE) was added the efficacy improved further. One study revealed that the speed of the contrast uptake by DCE MRI allows differentiating cancer from normal areas. There were wide variations in the specificity on MpMRI in different studies but sensitivity and NPV was high consistency and Fusion biopsy detected more clinical significant cancer than conventional systematic biopsy. One study with contrast enhanced TRUSB (on positive MRI) confirmed excellent sensitivity. MRI guided biopsy through transperineal route also improved clinically significant cancer detection but combined fusion and systematic biopsy had the highest detection rate. Fusion biopsy detects higher grade PCa than conventional biopsy. However, one of the studies found that fusion biopsy did not increase cancer detection rate and another one confirmed no added advantage of fusion biopsy over conventional biopsy, if overall outcome is cancer detection rate.

7.5 Discussions

The aim of this chapter was to examine the evidence that compares the accuracy of MpMRI with systematic (10–12 core) prostate biopsy, fusion biopsy (1–4 core), and final prostate pathology for prostate cancer diagnosis

Table 7.3 Result of the studies

Study	Aim	Study Type	(n)	Key Findings	(l)
[11]	To examine the performance of T2W and DW MRI after compare with histology	Prospective	199	T2W and DW MRI detects tumour in 92% cases	2-
[12]	The role of DCE MRI and MRSI for to detect PCa in biopsy negative men.	RCT	180	Combination of both this MRI offer 92% cancer detection rate	1-
[13]	Localisation of PCa by the speed of contrast uptake by DCE MRI	Prospective	30	Allows differentiate cancer and from remote areas	2-
[14]	To measure the diagnostic accuracy of MpMRI	Meta analysis	7 study	High specificity variable but high sensitivity and NPV	1-
[15]	MpMRI for accurate localization of tumour compared with histology	Prospective	75	T2W, DCE and DW MRI significantly detect PCa	2-
[16]	MRI guided biopsy can predict the aggressiveness of PCa	Prospective	518	DWI-Dbs had superior performance than MRS-Dbs in PZ	2-
[17]	Compare MRI guided biopsy and TRUS guided systemic biopsy	Prospective	132	Improves clinically significant cancer detection	2-
[18]	Accuracy of USS guided CE biopsy on +ve MRI but -ve biopsy patients.	Prospective	158	CE US targeted transrectal biopsy offers excellent sensitivity.	2-
[19]	Compare MRI guided biopsy and systemic biopsy through transperineal route	Prospective	182	Improves clinically significant cancer detection	2-
[20]	Compare MR-USS fusion biopsy with USS guided systemic biopsy	Prospective	1003	Increased detection of high risk PCa	2-
[21]	MR-USS fusion biopsy may better sample the true gland pahology	Prospective	582	32% higher detection of high risk PCa	2-
[22]	Compare MR-USS fusion biopsy with USS guided biopsy	Prospective	95	Improves detection of clinically significant cancer	2-
[23]	Compare MR-USS fusion biopsy with visual targeting biopsy	Prospective	125	Fusion biopsy did not increase cancer detection	2-
[24]	Compare MR-USS fusion biopsy with final prostate pathology	Prospective	54	Fusion biopsy detects more cancer	2-
[25]	Usefulness of MpMRI in detecting higher grade cancer compapre with fusion biopsy	Prospective	583	MpMRI is useful in high grade PCa	2-
[26]	Comparison MpMRI guided TB V systematic biopsies in the detection of PCa: a systematic literature review.	Systemic review	15 study	No advantage of TB but combined biopsy provide highest detection rate	1-

16 papers in total were reviewed and categorised into three groups-

1. Accuracy of MpMRI (Papers 1–4)
2. MpMRI compared with TRUS biopsy/ TPSB/ final histology (Papers 5–9)
3. MpMRI compared with fusion biopsy (Papers 10–16).

7.6   Limitations of this Review

These reviews also had some limitations-

1. Most of the studies did not perform power calculation prior to the study design, which raised the question for external validities for these studies.
2. In some studies, [1, 5] MpMRI was performed within 12 weeks of post biopsy; as it is known that haemorrhage after biopsy/ scarring can provide false positives on MRI and resolve within 12 weeks.
3. No study had performed cost analysis.
4. The PI-RADS scoring system was also not used by most of the studies for lesion characterisation.
5. Three studies [7, 9, 11] only included positive lesions on MpMRI but ignored negative scans.
6. In two studies [8, 16] authors confirmed a financial interest.

Despite limitations, the result on these studies has significant implications in clinical practice. Overall, MpMRI has a high efficacy in almost all studies and fusion biopsy is convenient for patients as fewer cores are taken to confirm the diagnosis. However, well designed controlled studies do not demonstrate a clear advantage of fusion biopsies over standard systematic biopsies in the primary setting as far as overall detection PCa is considered. However, fusion biopsy can detect more clinically significant cancer. For repeat biopsies, fusion biopsy is superior to standard systematic biopsies. The positive MpMRI and subsequent fusion biopsy could therefore be a possible solution to detect PCa in the scenarios of previous negative biopsies but ongoing suspicion of PCa.

7.7   Conclusions

This review reveals that MpMRI is a useful tool for PCa diagnosis prior to biopsy. In the repeat biopsy setting, image-targeted biopsies can detect more clinically significant prostate cancer in a positive MpMRI compared to standard systematic biopsies. However, only few studies have compared the results with saturation biopsies or with final histology.

In patients with a negative conventional TRUSB but ongoing suspicion of PCa, MpMRI can be a good guide for further management planning, however a negative MpMRI can't entirely exclude PCa. Fusion biopsy with fewer cores can detect more clinical significant cancer but can also miss some degree of clinically significant cancer and overall cancer detection rate was not higher than systemic biopsy, in many studies. In all studies, combined techniques detected most cancers (Standard and fusion biopsy) and with MpMRI (T2 + DW + DCE), the PCa detection rate was highest.

To recommend combined biopsy or only fusion biopsy in primary biopsy settings is debatable even if it may detect more and clinical significant cancers but no studies have performed cost analysis to recommend this in clinical practice. Additional larger randomised studies are required to compare two biopsy modalities to each other with the final prostatectomy specimen. A cost analysis needs to be performed to recommend this in routine clinical practice. Based on the findings of theses studies, future prospective PCa screening protocols are needed to evaluate the benefit of MpMRI as an independent modality, as well as MpMRI coupled with other screening parameters including tumour markers and measures of the PSA dynamics in detecting clinically significant cancers.

Appendix 1: 'Critical Appraisal Skills Programme' (CASP) Appraisal Tools

11 questions to help you make sense of case control study -.

How to use this appraisal tool -Three broad issues need to be considered when appraising a case control study:

● Are the results of the trial valid? ● What are the results ● Will the results help locally?

(Section A) (Section B) (Section C) The 11 questions on the following pages are designed to help you think about these issues systematically.The first two questions are screening questions and can be answered quickly. If the answer to both is "yes", it is worth proceeding with the remaining questions. There is some degree of overlap between the questions, you are asked to record a "yes", "no" or "can't tell" to most of the questions. A number of italicised prompts are given after each question. These are designed to remind you why the question is important. Record your reasons for your answers in the spaces provided.

(A) Are the results of the study valid?

Screening Questions

1. Did the study address a clearly focused issue? ☐Yes ☐Can't tell ☐No

HINT: A question can be focused in terms of The population studied, The risk factors studied, Whether the study tried to detect a beneficial or harmful effect?

2. Did the authors use an appropriate method to answer their question?

HINT: Consider ● Is a case control study an appropriate way of answering the question under the circumstances? (Is the outcome rare or harmful) ● Did it address the study question? ☐Yes ☐Can't tell ☐No.Is it worth continuing?

Detailed questions

3. Were the cases recruited in an acceptable way?

HINT: We are looking for selection bias, which might compromise validity of the findings

Are the cases defined precisely? Were the cases representative of a defined population? (Geographically and/or temporally?) Was there an established reliable system for selecting all the cases -Are they incident or prevalent?

Is there something special about the cases?

Is the time frame of the study relevant to disease/exposure? Was there a sufficient number of cases selected?Was there a power calculation?

4. Were the controls selected in an acceptable way?

HINT: We are looking for selection bias which might compromise The generalisibilty of the findings.Were the controls representative of defined population (geographically and/or temporally). Was there something special about the controls? Was the non-response high? Could non-respondents be different in any way? Are they matched, population based or randomly selected? Was there a sufficient number of controls selected?

5. Was the exposure accurately measured to ☐Yes minimise bias?

HINT: We are looking for measurement, recall or classification bias

Was the exposure clearly defined and accurately measured? Did the authors use subjective or objective measurements? Do the measures truly reflect what they are supposed to measure? (Have they been validated?) Were the measurement methods similar in the cases and controls? Did the study incorporate blinding where feasible? Is the temporal relation correct? (Does the exposure of interest precede the outcome?)

6. (a) What confounding factors have the List: authors accounted for? HINT: List the ones you think might be important, that The author missed. Genetic, Environmental, Socio-economic.

   (b) Have the authors taken account of the potential confounding factors in the design and/or in their analysis? HINT: Look for

• Restriction in design, and techniques e.g. modelling stratified-, regression-, or sensitivity analysis to correct, control or adjust for confounding factors

7. What are the results of this study? HINT: Consider -What are the bottom line results?Is the analysis appropriate to the design? How strong is the association between exposure and outcome (look at the odds ratio)? Are the results adjusted for confounding, and might confounding still explain the association? Has adjustment made a big difference to the OR?

   (B) What are the results?

8. How precise are the results? How precise is the estimate of risk?

HINT: Consider -Size of the P-value, Size of the confidence intervals, Have the authors considered all the important variables? How was the effect of subjects refusing to participate evaluated?

9. Do you believe the results? HINT: Consider -Big effect is hard to ignore! Can it be due to chance, bias or confounding? Are the design and methods of this study sufficiently flawed to make the results unreliable? Consider Bradford Hills criteria (e.g. time sequence, dose-response gradient, strength, biological plausibility)

    (C) Will the results help locally?

10. Can the results be applied to the local population? HINT: Consider whether -The subjects covered in the study could be sufficiently different from your population to cause concern. Your local setting is likely to differ much from that of the study. Can you quantify the local benefits and harms?

11. Do the results of this study fit with other available evidence? HINT: Consider all the available evidence from RCT's, systematic reviews, cohort studies and case-control studies as well for consistency.

One observational study rarely provides sufficiently robust evidence to recommend changes to clinical practice or within health policy decision-making. However, for certain questions observational studies provide the only evidence. Recommendations from observational studies are always stronger when supported by other evidence.

Appendix 2 – Grading of Recommendations

A new system for grading recommendations in evidence based guidelines Harbour, R. and Miller, J. 2001. The Scottish Intercollegiate Guidelines Network Grading Review Group. Scottish Intercollegiate Guidelines Network, Royal College of Physicians of Edinburgh, Edinburgh EH2 1JQ
    Levels of evidence-

1++ High quality meta-analyses, systematic reviews of RCTs, or RCTs with a very low risk of bias

1+ Well conducted meta-analyses, systematic reviews of RCTs, or RCTs with a low risk of bias

1+ Meta-analyses, systematic reviews or RCTs, or RCTs with a high risk of bias

2++ High quality systematic reviews of case-control or cohort studies *or* High quality case- control or cohort studies with a very low risk of confounding, bias, or chance and a high probability that the relationship is causal

2+ Well conducted case-control or cohort studies with a low risk of confounding, bias, or chance and a moderate probability that the relationship is causal

2+ Case-control or cohort studies with a high risk of confounding, bias, or chance and a significant risk that the relationship is not causal

3+ Non-analytic studies, e.g. case reports, case series

4+ Expert opinions.

Acknowledgement I would like to thank my advisor Mr. Nigel Parr, Consultant Urologist and Mr.Billy McWilliams, (Course Leader, MSc in Advanced Surgical Practice, Cardiff University, UK) who has also guided and supported me with his advice and suggestions throughout this review.

References

1. Arnold M, Karim-Kos HE, Coebergh JW, Byrnes G, Antilla A, Ferlay J et al (2015 Jun 30) Recent trends in incidence of five common cancers in 26 European countries since 1988: analysis of the European Cancer observatory. Eur J Cancer 51(9):1164–1187
2. Rais-Bahrami S, Siddiqui MM, Turkbey B, Stamatakis L, Logan J, Hoang AN et al (2013 Nov 30) Utility of multiparametric magnetic resonance imaging suspicion levels for detecting prostate cancer. J Urol 190(5):1721–1727
3. Applewhite JC, Matlaga BR, McCullough DL (2002 Aug 31) Results of the 5 region prostate biopsy method: the repeat biopsy population. J Urol 168(2):500–503
4. Igel TC, Knight MK, Young PR, Wehle MJ, Petrou SP, Broderick GA, Marino R, Parra RO (2001 May 31) Systematic transperineal ultrasound guided template biopsy of the prostate in patients at high risk. J Urol 165(5):1575–1579
5. Merrick GS, Taubenslag W, Andreini H, Brammer S, Butler WM, Adamovich E et al (2008 Jun 1) The morbidity of transperineal template-guided prostate mapping biopsy. BJU international 101(12):1524–1529
6. Bonekamp D, Jacobs MA, El-Khouli R, Stoianovici D, Macura KJ (2011 May 4) Advancements in MR imaging of the prostate: from diagnosis to interventions. Radiographics 31(3):677–703
7. Isebaert S, Van den Bergh L, Haustermans K, Joniau S, Lerut E, De Wever L, De Keyzer F, Budiharto T, Slagmolen P, Van Poppel H, Oyen R (2013 Jun 1) Multiparametric MRI for prostate cancer localization in correlation to whole-mount histopathology. J Magn Reson Imaging 37(6):1392–1401
8. Marks L, Young S, Natarajan S (2013 Jan) MRI–ultrasound fusion for guidance of targeted prostate biopsy. Curr Opin Urol 23(1):43–50
9. Hambrock T, Hoeks C, Hulsbergen-van de Kaa C, Scheenen T, Fütterer J, Bouwense S et al (2012 Jan 31) Prospective assessment of prostate cancer aggressiveness using 3-T diffusion-weighted magnetic resonance imaging–guided biopsies versus a systematic 10-core transrectal ultrasound prostate biopsy cohort. Eur Urol 61(1):177–184
10. Sackett DL (1997) Evidence-based medicine: how to practice and teach EBM. WB Saunders Company
11. Rud E, Klotz D, Rennesund K, Baco E, Berge V, Lien D et al (2014 Dec 1) Detection of the index tumour and tumour volume in prostate cancer using T2-weighted and diffusion-weighted magnetic resonance imaging (MRI) alone. BJU Int 114(6b):E32–E42
12. Sciarra A, Panebianco V, Ciccariello M, Salciccia S, Cattarino S, Lisi D et al (2010 Mar 15) Value of magnetic resonance spectroscopy imaging and dynamic contrast-enhanced imaging for detecting prostate cancer foci in men with prior negative biopsy. Clin Cancer Res 16(6):1875–1883
13. Valentini AL, Gui B, Cina A, Pinto F, Totaro A, Pierconti F et al (2012 Nov 30) T2-weighted hypointense lesions within prostate gland: differential diagnosis using wash-in rate parameter on the basis of dynamic contrast-enhanced magnetic resonance imaging—Hystopatology correlations. Eur J Radiol 81(11):3090–3095
14. de Rooij M, Hamoen EH, Fütterer JJ, Barentsz JO, Rovers MM (2014) Accuracy of multiparametric MRI for prostate cancer detection: a meta-analysis. Am J Roentgenol 202(2):343–351
15. Isebaert S, Van den Bergh L, Haustermans K, Joniau S, Lerut E, De Wever L et al (2013 Jun 1) Multiparametric MRI for prostate cancer localization in correlation to whole-mount histopathology. J Magn Reson Imaging 37(6):1392–1401

16. Zhang J, Xiu J, Dong Y, Wang M, Han X, Qin Y et al (2014 May 1) Magnetic resonance imaging-directed biopsy improves the prediction of prostate cancer aggressiveness compared with a 12-core transrectal ultrasound-guided prostate biopsy. Mol Med Rep 9(5):1989–1997

17. Quentin M, Blondin D, Arsov C, Schimmöller L, Hiester A, Godehardt E et al (2014 Nov 30) Prospective evaluation of magnetic resonance imaging guided in-bore prostate biopsy versus systematic transrectal ultrasound guided prostate biopsy in biopsy naïve men with elevated prostate specific antigen. J Urol 192(5):1374–1379

18. Cornelis F, Rigou G, Le Bras Y, Coutouly X, Hubrecht R, Yacoub M, Pasticier G, Robert G, Grenier N (2013 Oct) Real-time contrast-enhanced transrectal US-guided prostate biopsy: diagnostic accuracy in men with previously negative biopsy results and positive MR imaging findings. Radiology 269(1):159–166

19. Kasivisvanathan V, Dufour R, Moore CM, Ahmed HU, Abd-Alazeez M, Charman SC et al (2013 Mar 31) Transperineal magnetic resonance image targeted prostate biopsy versus transperineal template prostate biopsy in the detection of clinically significant prostate cancer. J Urol 189(3):860–866

20. Siddiqui MM, Rais-Bahrami S, Turkbey B, George AK, Rothwax J, Shakir N et al (2015 Jan 27) Comparison of MR/ultrasound fusion–guided biopsy with ultrasound-guided biopsy for the diagnosis of prostate cancer. JAMA 313(4):390–397

21. Siddiqui MM, Rais-Bahrami S, Truong H, Stamatakis L, Vourganti S, Nix J et al (2013 Nov 30) Magnetic resonance imaging/ultrasound–fusion biopsy significantly upgrades prostate cancer versus systematic 12-core transrectal ultrasound biopsy. Eur Urol 64(5):713–719

22. Puech P, Rouvière O, Renard-Penna R, Villers A, Devos P, Colombel M et al (2013 Aug) Prostate cancer diagnosis: multiparametric MR-targeted biopsy with cognitive and transrectal US–MR fusion guidance versus systematic biopsy—prospective multicenter study. Radiology 268(2):461–469

23. Wysock JS, Rosenkrantz AB, Huang WC, Stifelman MD, Lepor H, Deng FM et al (2014 Aug 31) A prospective, blinded comparison of magnetic resonance (MR) imaging–ultrasound fusion and visual estimation in the performance of MR-targeted prostate biopsy: the PROFUS trial. Eur Urol 66(2):343–351

24. Le JD, Stephenson S, Brugger M, Lu DY, Lieu P, Sonn GA et al (2014 Nov 30) Magnetic resonance imaging-ultrasound fusion biopsy for prediction of final prostate pathology. J Urol 192(5):1367–1373

25. Rais-Bahrami S, Siddiqui MM, Turkbey B, Stamatakis L, Logan J, Hoang AN, Walton-Diaz A, Vourganti S, Truong H, Kruecker J, Merino MJ (2013 Nov 30) Utility of multiparametric magnetic resonance imaging suspicion levels for detecting prostate cancer. J Urol 190(5):1721–1727

26. van Hove A, Savoie PH, Maurin C, Brunelle S, Gravis G, Salem N et al (2014 Aug 1) Comparison of image-guided targeted biopsies versus systematic randomized biopsies in the detection of prostate cancer: a systematic literature review of well-designed studies. World J Urol 32(4):847–845

Chapter 8
A Geneticist's View of Prostate Cancer: Prostate Cancer Treatment Considerations

Abraham Eisenstark

Abstract Prostate cancer remains a life-threatening disease of men. While early detection has been helpful to reduce the mortality rate, we currently do not have a desired therapy. In recent years, new strategies have been proposed to treat prostate cancers with poor prognosis by utilizing genetically modified bacteria, including *Salmonella typhimurium* that preferentially replicate within solid tumors (1000:1 and up to 10,000:1 compared to non-cancerous tissue) destroying cancer cells without causing septic shock that is typically associated with wild-type *S. typhimurium* infections. Furthermore, these bacteria have the potential to be utilized as drug delivery systems to more effectively target different subpopulations of prostate tumor cells. This chapter reviews progress in using genetically modified *S. typhimurium* for destruction of prostate tumors.

Keywords Prostate cancer · Cancer therapy · Genetically modified bacteria · Drug delivery · Prostate cancer subpopulations

8.1 Introduction

This chapter reviews progress in using genetically modified *S. typhimurium* for destruction of prostate cancer cells in culture and in solid cancer tissue [1]. It discusses the potential and future prospects for applications in clinical trials as novel prostate cancer therapy for advanced stages of the disease. We further discuss potential combinational therapies for optimal destruction of prostate cancer cells.

Research Phases PHASE 1. Starting with a desire to understand prostate tumor development and therapy, our research lab delved into the study of near-ultraviolet radiation causing oxidative damage and DNA repair that occurred. This research

A. Eisenstark (✉)
Cancer Research Center, University of Missouri, Columbia, MO, USA
e-mail: eisenstarka@missouri.edu

© Springer International Publishing AG, part of Springer Nature 2018
H. Schatten (ed.), *Cell & Molecular Biology of Prostate Cancer*,
Advances in Experimental Medicine and Biology 1095,
https://doi.org/10.1007/978-3-319-95693-0_8

phase involved the construction of over 200 mutants of *S. typhiurium* subpopulations. **PHASE 2.** The next step was to demonstrate that the selected *S. typhimurium* mutant 2631 had anti-tumor effects. **PHASE 3.** This phase was to demonstrate that 2631 actually limits prostate tumor progression in mice. **PHASE 4.** Suggests experimental approaches to dealing with cancer in human populations.

History of Mutant Collection Regarding the ancestry of our therapeutic mutant *Salmonella typhimurim* 2631, it originated from strains of *Salmonella typhimurium* isolates from sewers of Scandinavia by Kaare Lilleengen, designated as LT1 thru LT22 [2]. Later, Professor Joshua Lederberg arranged, via the Embassy in Copenhagen, to receive a set of these LT strains [3]. (Note: Nobelist Lederberg and Esther Lederberg developed the early molecular genetic technology in the construction of cancer therapeutic mutant 2631). They were then deposited at the *S. typhimurium* collection at Carnegie Institute of Genetics at Cold Spring Harbor, Long Island, N.Y. directed by Miloslav Demerec [4]. Upon the death of Dr. Demerec, the curatorship of Salmonella was shifted to Professor Kenneth Sanderson, University of Calgary, Canada [5]. However, since these mutants (several hundred in number) had also been stored in sealed nutrient agar stabs in quintuplicates, replicas of each were sent to Phillip Hartman, Johns Hopkins University, and Professor Abraham Eisenstark, Kansas State University, and the collection is now at the Cancer Research Center, Columbia, Missouri. Cultures have also been stored in lyophilized sealed tubes, and freeze-dried and in refrigerated frozen nutrient broth, distributed in several labs, worldwide.

Until recently, the standard method of dealing with prostate tumors has been radiation, surgery and drugs. There is currently a shift to genetic approaches. This is apparent from prescriptions of newer drugs, from the literature being distributed by the pharmaceutical industries, by all public press sources, as well as through a greater knowledge by individuals on the aging of all biological entities, and in scientific publications including the previous chapters in this volume. From scientific researchers, there have been some important technical introductions. Perhaps the outstanding innovation has been the ability to base sequence DNA samples, in large batches, quickly, cheaply, and to annotate the gene product. This has progressed in four phases.

Phase 1. Oxidative Damage by Near–Ultraviolet Radiation, and Irradiated Byproducts To obtain our final therapeutic CRC2631 anti prostate tumor strains, these mutagenic agents were used, and standard methods of selection were employed [6–9].

Phase 2. S. Typhimurium as Anti–Prostate Agent The therapeutic *Salmonella* strain CRC2631 target prostate tumors that arise in the transgenic adenocarcinoma mouse prostate (TRAMP) model (C57BL/6-Tg(TRAMP)8247Ng/j). Weekly

CRC2631 intraperitoneal injections into TRAMP transgenic males significantly reduced tumor size, but also inhibit tumor progression. This study opens the door for testing *Salmonella*-based monotherapies for treatment of cancer in human clinical trials. Also, combining such treatments with other therapies could improve the outcome of prostate cancer.

Hemizygous TRAMP males manifest progressive prostate tumors that begin as early prostatic intraepithelial neoplasia (PIN) and can progress to differentiated adenocarcinoma. Moreover, the tumors that originate in the prostate can metastasize to distant sites.

Phase 3. Our Cancer Research Center Involvement of New Therapeutic 2631 Strain of *S. typhimurium*   Further research involves not only critical testing *in vivo* with known tumor strains and tumor-infested mice, but our Cancer Research Center is now arranging formal approval, and an appropriate protocol with our medical school hospital, for clinical trials in human volunteers [10–17].

Phase 4   As described in the records of the National Institute of Medicine, there is comprehensive literature on genetic aspects of prostate tumor disease and therapy. There are vast differences in prostate tumor occurrence depending on geographic, environmental, human genetic, and nutritional factors. Also, important sets of information have resulted from the examination of the records of various groups, such as the Mormons, masons, and Ashkenazy Jews of Central Europe, and people from many other cultures throughout the world.

A number of questions are raised that need attention to reduce the incidence of prostate tumors.

Thanks to technical advances of DNA base sequencing and the ability to visualize the tracking of Salmonella nanoparticles as they invade and alter structures of tumors, clearer insights are emerging into the prostate cancer effects. It is now possible to identify the genes of the therapeutic *Salmonella* 2631 strain, the DNA of the tumor target and the DNA of the cancer victim. To date, while the sequence of therapeutic agents (and the coded nanoparticles) may be identified, those of the tumor targets and pertinent human tumor-associated genes have yet to be identified.

Since key goals in dealing with prostate cancer is to reduce the incidence of cases and to raise years of longer life expectancy, a challenge for genetic researchers is to identify the male human genes that may be involved. While there may remain factors of day-today living social qualities, there are many questions to be addressed, such as:

Why are incidence and longevity rates for African-Americans more frequent? Are there *also differences among other ethnic* groups? Among brothers and father-sons in the same family? Between twins?

8.2 Conclusions and Further Reading

The use of other genetically modified (attenuated) bacteria as therapy for advanced and metastatic breast cancer has previously been reviewed by our group [1] and several other groups have contributed new knowledge for the use of genetically modified bacteria particularly for new therapies of advanced cancers for which commonly used therapies have become ineffective [3, 18–22]. This line of research for prostate cancer has high potential to determine therapies to overcome the current limitations offered by surgery, hormone ablation and chemotherapies.

Acknowledgements Thanks to Jacki Kian Mehr for her professional skills in searching pertinent science literature, and in preparation of this chapter. I also appreciate my association with students and distinguished scientists worldwide for their insights into the role of genetics in cancer disease.

References

1. Kazmierczak R, Choe E, Sinclair J, Eisenstark A (2015) Direct attachment of nanoparticle cargo to Salmonella typhimurium membranes designed for combination bacteriotherapy against tumors. Methods Mol Biol 1225:151–163. https://doi.org/10.1007/978-1-4939-1625-2_11
2. Lilleengen K (1948) Typing of Salmonella Typhimurium by means of bacteriophage. Acta Pathol Microbiol Scand.77 (supple.): 125
3. Bermudes D, Low KB, Pawelek J, Feng M, Belcourt M, Zheng LM, King I (2001) Tumour-selective Salmonella-based cancer therapy. Biotechnol Genet Eng Rev 18:219–233 Review
4. Demerec M (1958) Genetic structure of the Salmonella cromosome. X Int Congr Genet 1:55–62
5. Lederberg J (2000 May) The dawning of molecular genetics. Trends Microbiol 8(5):194–195
6. Eisenstark A (1971) Mutagenic and lethal effects of visible and near-ultraviolet light on bacterial cells. Adv Genet 16:167–198 Review
7. Ferron WL, Eisenstark A, Mackay D (1972 Sep 14) Distinction between far- and near-ultraviolet light killing of recombinationless (recA) Salmonella typhimurium. Biochim Biophys Acta 277(3):651–658
8. McCormick JP, Fischer JR, Pachlatko JP, Eisenstark A (1976 Feb 6) Characterization of a cell-lethal product from the photooxidation of tryptophan: hydrogen peroxide. Science 191(4226):468–469
9. Ananthaswamy HN, Eisenstark A (1977 Apr) Repair of hydrogen peroxide-induced single-strand breaks in Escherichia coli deoxyribonucleic acid. J Bacteriol 130(1):187–191
10. Ahmad SI, Eisenstark A (1979 May 4) Thymidine sensitivity of certain strains of Escherichia coli K12. Mol Gen Genet 172(2):229–231
11. Eisenstark A (1989) Bacterial genes involved in response to near-ultraviolet radiation. Adv Genet 26:99–147 Review
12. Arber W (2000 Jan) Genetic variation: molecular mechanisms and impact on microbial evolution. FEMS Microbiol Rev 24(1):1–7
13. Edwards K, Linetsky I, Hueser C, Eisenstark A (2001 May 30) Genetic variability among archival cultures of Salmonella typhimurium. FEMS Microbiol Lett 199(2):215–219
14. Faure D, Frederick R, Włoch D, Portier P, Blot M, Adams J (2004 Oct) Genomic changes arising in long-term stab cultures of Escherichia coli. J Bacteriol 186(19):6437–6442

15. Porwollik S, Wong RM, Helm RA, Edwards KK, Calcutt M, Eisenstark A, McClelland M (2004 Mar) DNA amplification and rearrangements in archival Salmonella enterica serovar typhimurium LT2 cultures. J Bacteriol 186(6):1678–1682
16. McClelland M, Sanderson KE, Spieth J, Clifton SW, Latreille P, Courtney L, Porwollik S, Ali J, Dante M, Du F, Hou S, Layman D, Leonard S, Nguyen C, Scott K, Holmes A, Grewal N, Mulvaney E, Ryan E, Sun H, Florea L, Miller W, Stoneking T, Nhan M, Waterston R, Wilson RK (2001 Oct 25) Complete genome sequence of Salmonella enterica serovar Typhimurium LT2. Nature 413(6858):852–856
17. Chakrabarty AM (2003 May) Microorganisms and cancer: quest for a therapy. J Bacteriol 185(9):2683–2686 Review
18. Loeffler M, Le'Negrate G, Krajewska M, Reed JC (2007 Jul 31) Attenuated Salmonella engineered to produce human cytokine LIGHT inhibit tumor growth. Proc Natl Acad Sci U S A 104(31):12879–12883
19. Fensterle J, Bergmann B, Yone CL, Hotz C, Meyer SR, Spreng S, Goebel W, Rapp UR, Gentschev I (2008 Feb) Cancer immunotherapy based on recombinant Salmonella enterica serovar typhimurium aroA strains secreting prostate-specific antigen and cholera to xin subunit B. Cancer Gene Ther 15(2):85–93
20. Forbes NS, Munn LL, Fukumura D, Jain RK (2003 Sep 1) Sparse initial entrapment of systemically injected Salmonella typhimurium leads to heterogeneous accumulation within tumors. Cancer Res 63(17):5188–5193
21. Low KB, Ittensohn M, Le T, Platt J, Sodi S, Amoss M, Ash O, Carmichael E, Chakraborty A, Fischer J, Lin SL, Luo X, Miller SI, Zheng L, King I, Pawelek JM, Bermudes D (1999) Lipid a mutant Salmonella with suppressed virulence and TNFα induction retain tumor targeting in vivo. Nat Biotechnol 17:37–41
22. Forbes NS (2010 Nov) Engineering the perfect (bacterial) cancer therapy. Nat Rev Cancer 10(11):785–794. https://doi.org/10.1038/nrc2934 Epub 2010 Oct 14. Review

Printed in the United States
By Bookmasters